1

A Comparative Review of Interaction and Nonlinear Modeling

Edward E. Rigdon
Georgia State University

Randall E. Schumacker
University of North Texas

Werner Wothke
SmallWaters Corporation, Chicago, Illinois

Interaction effects are common in the real world. A business manager knows that a dollar spent on advertising will yield a greater marginal return in sales for a product with wide distribution than for a product with narrow distribution (Lilien & Kotler, 1983). In contrast, an interaction may dampen the individual effects of the two variables, as when two noises combine to create a zone of apparent quiet. Two gases may be relatively harmless when released into the atmosphere separately, but may yield lethal toxins when released together. Smith and Sasaki (1979) described a "consistency" interaction effect. When two variables both deviate from their means in the same direction, their joint effect is enhanced, but if the deviations are in opposite directions, then the interaction effect is opposite, as well.

Techniques for representing and testing interaction effects are familiar in the regression and analysis-of-variance methodologies. In structural equation modeling (SEM), researchers have a choice of two basic techniques. If one or both of the interacting variables is discrete, or can be made so, researchers can apply a "multisample" approach, with the interaction effects becoming apparent as differences in parameter estimates when the same model is applied to different but related sets of data. If both interacting variables are continuous, researchers can apply Kenny and Judd's (1984) indicant product analysis procedure for modeling the interaction in a single sample, although this procedure can lead to a number of practical and statistical problems. Because of those problems, researchers have typically either avoided modeling interactions, or else adapted their data to the multisample approach. Recent developments in software, and contributions by Jöreskog and Yang (1996) and Ping

(1995), have made it much easier to apply the Kenny and Judd approach, and have provided a fresh look at the statistical issues involved. Consequently, researchers have renewed their interest in the indicant product procedure in SEM.

Unfortunately, the excitement over these developments may have obscured the real relative merits of the multisample and indicant product approaches, in much the same way that earlier excitement over SEM methods in general led some researchers to ignore basic issues (Cliff, 1983). Thus, this may be a good time for a critical comparison of the multisample and indicant product approaches. A careful review suggests that the multisample approach is still the most useful procedure for modeling latent variable interaction effects, under the widest set of circumstances, whereas the indicant product approach should be reserved for particular situations where it will yield superior results.

REVIEW

In general, an interaction effect exists whenever the character of the relationship between variables A and B is affected by the level of variable C (Keppel & Zedeck, 1989). We can demonstrate this by examining a multiple regression equation without interactions:

$$A = \beta_1 * B + \beta_2 * C + e \quad (1)$$

The first partial derivatives of this equation with respect to B and C describe how a change in the level of one predictor affects the level of A, when the other predictor is held constant. These partial derivatives are, respectively, β_1 and β_2. They show, for example, that the only effect of a change in B on variable A is the "main" effect through β_1. By contrast, consider a regression equation with an interaction term:

$$A = \beta_1 * B + \beta_2 * C + \beta_3 * (B * C) + e \quad (2)$$

Now the first partial derivatives are $\beta_1 + \beta_3 * C$ and $\beta_2 + \beta_3 * B$. In other words, a change in B now affects A not only through B's main effect but also as a function of the level of C. Accordingly, the main effect of B on A, β_1, can be thought of as an "average" value computed over the range of values for variable C that are represented in the data set (Keppel & Zedeck, 1989).

When researchers suspect an interaction relationship, they may be tempted to avoid SEM altogether, in favor of older techniques such as regression or analysis of variance. The older techniques provide researchers not only with well-established, formal procedures for dealing with interactions, but also with a body of practical experience. These techniques also provide an apparent simplification, because they substantially reduce the number of choices that researchers must make. The disadvantages of this approach, however, are well known. If variables are measured with error, parameter estimates may be biased. Adopting these techniques may reduce the level of rigor in the statistical analysis. Researchers may also find that they need the flexibility of SEM methods to deal with other aspects of their model, even at the cost of substantially increased complexity in dealing with the interaction effect.

The Indicant Product Approach

Kenny and Judd's (1984) indicant product approach to latent variable interactions involves modeling an "interaction construct," say ξ_3, which is a function of the "main effect constructs," ξ_1 and ξ_2. The structural model has the form:

$$\eta_1 = \beta_1 * \xi_1 + \beta_2 * \xi_2 + \beta_3 * \xi_3 + \zeta_1 \quad (3)$$

where η_1 is the dependent construct and ζ_1 is an error term. The relationship between η_1 and ξ_3 is itself linear. The researcher creates ξ_3 so as to represent the kind of nonlinear relationship believed to exist between η_1 and ξ_1 and ξ_2. For example, to model a multiplicative interaction between ξ_1 and ξ_2, ξ_3 must be made equal to $\xi_1 * \xi_2$. Measures or indicators of ξ_3 are then created as functions of the measures of ξ_1 and ξ_2, in a similar way. Kenny and Judd (1984), Hayduk (1987), and Bollen (1989) have all described the procedure for multiplicative interactions in some detail. It is important to note that the method relies on an external indicator of the hypothesized interaction for model identification.

The Multisample Approach

In the multisample approach, the different samples are defined by the different levels of one or both of the interacting variables. If interaction effects are present, then certain parameters should have different values in different samples (Bagozzi & Yi, 1989, offered a treatment). For example, in the context of an experiment, MacKenzie and Spreng (1992) tested an interaction model of advertising effectiveness. Their model proposed, in part, that a person's attitude toward an advertisement (A_{ad}) interacts with the person's motivation to process the ad (Motivation) to affect the person's attitude toward the advertised brand (A_{brand}) (see Figure 1.1). The general structural model may be represented as:

$$A_{brand} = \alpha + \beta * A_{ad} + \zeta \quad (4)$$

where α is the intercept and ζ is an error term.

MacKenzie and Spreng (1992) hypothesized that subjects in the high-Motivation group would have a higher mean level of A_{brand} after viewing the

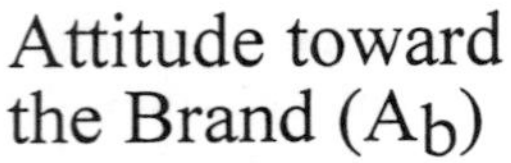

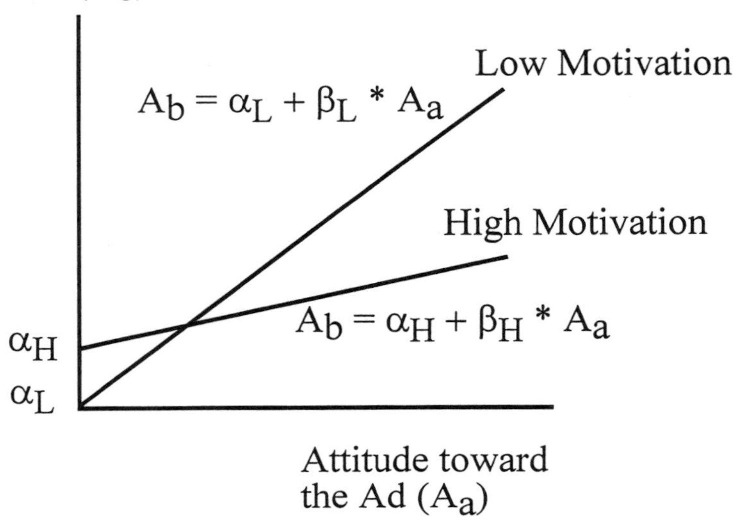

FIG. 1.1. Motivation moderating the intercept and slope of the regression relationship between Attitude toward the Ad and Attitude toward the Brand. ("H" indicates high Motivation and "L" indicates low Motivation.)

ad. This would be reflected in a higher intercept term for the subjects in this condition. They proposed that A_{ad} should have a positive "main" effect (as represented by β) on A_{brand}. However, they further proposed that the interaction between A_{ad} and Motivation would make the A_{ad}-A_{brand} relationship *weaker* in the high-Motivation condition (see Fig. 1.1).

Under the multisample approach, researchers investigate interaction effects using chi-square difference tests. Researchers first estimate a model where the parameters in question are constrained to be equal across the groups, and then estimate a model where the parameters are allowed to differ in the two samples. A significant chi-square difference suggests that the equality constraints are not consistent with the data, and thus that an interaction effect exists.

COMPARATIVE REVIEW

Data Types

A comparative review of the indicant product and multisample approaches should begin with the obvious. The indicant product approach is the "natural" approach when both interacting variables are continuous, whereas the multisample approach is the logical choice when one or both of the interacting variables are *discrete* or *categorical*. The discrete variable may be an experimental condition or a demographic variable such as gender. To apply the traditional multisample approach to the continuous-continuous case, one or both of the continuous variables must be reduced to a nominal or ordinal scale. Such a transformation produces an immediate loss of information, which is undesirable (Russell & Bobko, 1992). The transformation is especially problematic if the interacting variables are measured with error or if they are latent rather than observed. The presence of measurement error raises the possibility that cases will be misclassified, leading to biased parameter estimates for one or more subsamples. With multiple measures of the latent variables, researchers will naturally think of using factor scores to classify cases. However, factor scores are also problematic, due to their well-known indeterminacy (Bollen, 1989). Researchers might turn to other methods for assigning cases to groups, such as discriminant analysis, or perhaps to a combination of methods for increased confidence, but published applications of this approach are scarce.

Even after assigning cases to a continuum, researchers would still have to decide where to divide the continuum into categories. Ideally, the division(s) would reflect prior knowledge regarding the range of the categorized variable where the interaction effect is expected to occur. By reflex, researchers may choose division points that yield equal sample sizes in the various groups, but naive application of such a rule may actually obscure the interaction relationship.

Continuous moderators can be used as parameters in the model so that each case (or subject) has its own structural equation model. This type of analysis can be done only with raw data, which are assumed to be multivariate normal, conditional on the value of the moderator. Though this solves the problems of loss of information and the arbitrary choice of where to subdivide the continuous distribution, at least three problems—or questions—remain. First, having a different model for each case can dramatically increase computer time required to fit the model, particularly with large sample sizes, because there are N different predicted covariance matrices to invert to find the likelihood function. Second, when every subject has a model that depends on the data for that subject, there is an increased chance that an improbable model will be generated. For example, a non-positive definite covariance matrix might be generated given the specified model (Wothke, 1993). Such problems make working with interaction in this way more sensitive to starting values than usual. Third, the presence of measurement error in the interacting variable may have biasing effects the same as it does in the multisample or multiple regression approaches.

When one or more of the interacting variables is measured on an ordinal scale, the indicant product approach may force the researcher to confront the problems involved in dealing with polychotomous variables in SEM (see, e.g., Lee, Poon, & Bentler, 1992; Rigdon & Ferguson, 1991). If, as is common, the ordinal variable is taken to be a discrete reflection of a continuous latent variable, then researchers are normally advised to utilize polychoric or polyserial correlations. Pearson product-moment correlations may be substantially biased (Ollson, 1979) and the scale of the sample covariance becomes arbitrary. The use of polychoric or polyserial correlations leads to complications in assessing the fit of structural equation models, as researchers may be forced to adopt asymptotically distribution free (ADF) methods, which require *very* large samples (Muthén & Kaplan, 1992; although bootstrapping appears to offer a less demanding alternative—see Yung & Bentler, 1994). Furthermore, both the indicant product and multisample approaches require, as input, raw data or covariances rather than correlations. So there is a strong disincentive toward applying the indicant product approach with ordinal data.

Sample Size and Parsimony

The indicant product approach explicitly avoids the sample splitting required in the multisample approach. In a methodology that relies so heavily on asymptotic properties and where those properties may only be achieved with relatively large sample sizes, this is a serious limitation. Furthermore, this reduction in sample size may confound results from chi-square difference tests. MacCallum, Roznowski, and Necowitz (1992) noted a substantial degree of instability in fit indices in even moderately sized samples. Thus, it is possible that the multisample approach may yield subsamples that are so small that instability in the chi-square statistic will mislead the researcher into believing that an interaction effect exists, whether it does or not. Preserving sample size is even more important for researchers who want to assess model fit with cross-validation techniques, some of which require additional sample splitting.

In general, the indicant product approach may be somewhat more parsimonious than the multisample approach, although the difference may be small. In comparing their indicant product model with an interaction to an indicant product model with only main effects, Kenny and Judd (1984) showed that the interaction model includes only one additional free parameter—the parameter linking the interaction construct to the dependent construct. All other additional parameters associated with the interaction effect are exact functions of parameters included in the main effects model. The multisample approach potentially involves the estimation of many additional parameters. The researcher could estimate one full set of measurement and structural parameters for each group in the analysis. This practice, however, is often unjustified. Why should measurement parameters, for example, differ across groups, just because there is a structural interaction? MacKenzie and

Spreng (1992) made the point that imposing equality constraints across groups for measurement parameters ensures that the constructs in one group are substantively equal to the constructs in the other groups. Such constraints lead to a very parsimonious model, although researchers will need a certain minimal sample size within each group before they can expect the constraints to hold in practice. In addition, because one of the interacting constructs is used to define the groups themselves, the researcher will not estimate *any* parameters for that construct, in any subsample. In net terms, therefore, the difference in parsimony between the indicant product approach and the multisample approach may be small; and a parsimonious specification may partially ameliorate the negative effects of dividing the sample among groups.

Practical Issues

The indicant product interaction model developed by Kenny and Judd (1984) required the researcher to establish nonlinear constraints among different model parameters. At that time, COSAN (Fraser, 1980) was virtually the only SEM package that allowed researchers to specify nonlinear constraints directly. Researchers could also have used Rindskopf's (1984) "phantom variable" technique to achieve the same end, as illustrated by Hayduk (1987) and Bollen (1989), but that procedure was particularly onerous. Currently, Mx (Neale, 1995), LISREL 8 (Jöreskog & Sörbom, 1993), SAS PROC CALIS (Hartmann, 1995), and COSAN offer direct specification of nonlinear constraints, while other popular packages such as AMOS (Arbuckle, 1995) and EQS (Bentler, 1993) currently do not. By contrast, almost every popular SEM package offers multisample analysis. Even with the developments in modeling software, however, specifying an indicant product interaction model is still a tedious, complex undertaking, with much room for careless errors.

In addition, researchers who insist on having many measures of their constructs (in order to provide a strong test of their measurement model) will find the indicant product approach unwieldy. If constructs A and B have a and b measures, respectively, then the interaction construct AB will have $a * b$ measures. If each construct has four measures, as is necessary to apply Anderson and Gerbing's (1988) internal consistency criterion, then the interaction construct will have *16* measures. Counting the original four measures of the two "main effect" constructs, and at least one measure of the dependent construct, the model will have at least 25 manifest variables before considering any other constructs. This complexity can lead to convergence problems when it comes to actually estimating the model.

Ping (1995) has examined techniques for minimizing these practical problems. Ping's intuitive conclusion is that, if the measures of each main effect construct are sufficiently unidimensional, then researchers can adopt any one of several shortcut procedures to minimize these practical difficulties. For example, instead of creating 16 measures of the interaction construct, researchers might simply create *1*, by computing the product of one measure from each of the main effect constructs. If the original measures were truly unidimensional, then the product of any pair should point to the same interaction construct as the product of any other pair. In such cases, Ping also suggested a two-step process of first estimating the main effects only, then *fixing*, rather than constraining, the interaction parameters at the values implied by Kenny and Judd's (1984) model, and then estimating the interaction effect. This procedure allows researchers to estimate the interaction effect—the parameter linking the interaction construct with the dependent construct—without actually including nonlinear constraints in the model. When Ping's shortcuts are valid, they promise to make specifying indicant product interaction models nearly as easy as specifying multisample interaction models.

Flexibility

One important limitation of the indicant product approach is that it is only well defined for multiplicative interactions. By contrast, the multisample approach can be applied to most any type of interaction relationship. Jaccard and Wan (1996) demonstrated an extension of this approach to higher-order interactions. Furthermore, in order to apply the indicant product approach, researchers must specify a priori the particular functional form of the interaction, so that they can determine how to construct measures of the interaction. In extending the indicant product approach to other forms of interaction, the researcher may be forced to develop significant statistical theory. In addition, the multisample approach makes it easy to incorporate mean effects in the model, along with the interaction effect, whereas some latent mean parameters may not be identified in the single-sample indicant product approach. Jöreskog and Yang's (1996) treatment of the indicant product approach includes mean terms for some of the measures as essential components, but their statistical development points out additional problems, as is discussed later.

Multicollinearity and Distributional Problems

Applying the indicant product approach may lead to substantial multicollinearity, precisely because the measures associated with the interaction construct are functions of the measures of the main effect constructs. The computed measures that result may be highly correlated with the original measures. Besides the usual problems associated with multicollinearity, it can be particularly devastating in factor analytic measurement models, causing measures to be more highly correlated with measures of other constructs than they are with measures of the same construct. Smith and Sasaki (1979) pointed out that, if the interaction relationship is multiplicative and if the original measures are basically symmetric, then researchers can avoid multicollinearity by centering the measures (adjusting their means to zero) before computing the product terms. However, this adjustment turns the multiplicative interaction into Smith and Sasaki's consistency effect, which is functionally different from a straightforward multiplicative interaction. Furthermore, the centering technique does not necessarily solve the multicollinearity problem for any *other* type of interaction.

The interaction model also presents distributional problems that are more serious than those associated with SEM techniques in general. First, Kenny and Judd's (1984) statistical development of the multiplicative interaction case itself relies on the assumption that the original measures are normally distributed. They make heavy use of a result regarding the variance of the product of two normal variables. If the original measures are severely non-normal, the variance of the product variable can be very different from the value implied by their development, and the interaction model may perform poorly. Of course, transformation may result in suitably normal distributions for the original variables. Furthermore, Jöreskog (personal communication, February 14, 1995) indicated that researchers can deal with non-normality merely by relaxing some of the constraints that are part of the Kenny–Judd interaction model. Thus, the cost of non-normality here may be no more than a loss of parsimony.

The computed measures of the interaction construct may also precipitate other potentially important distributional problems. Even if the original measures are normally distributed, the new computed measures often will not be—in the multiplicative case, certainly not. This violation of the distributional assumption underlying maximum likelihood estimation may lead researchers using the indicant product approach to turn to ADF methods. Unfortunately, as Jöreskog and Yang (1996) pointed out, in the case of multiplicative interactions, standard distribution-free methods may be unavailable, because moments of the product measures may be functions of the moments of the original measures. Thus, the asymptotic weight matrix associated with the covariance matrix for an interaction model may be singular in the population. Distribution-free estimation methods that use the inverse of that weight matrix will therefore fail. Jöreskog and Yang recommended an augmented weight matrix as a theoretically correct but practically difficult solution. Researchers may also attempt to circumvent this problem by estimating the model using a distribution-sensitive method but evaluating the results with a bootstrapping technique. The rescaled bootstrapping method outlined by Bollen and Stine (1993) may be particularly appropriate in this case. Note that techniques for computing bootstrap estimates of the weight matrix itself (Yung & Bentler, 1994) will not be helpful in this case, because the weight matrix is singular in the population. The multisample approach, by contrast, avoids all of these difficulties. In fact, the researcher may be able to preserve linear relations within each group.

Summary

What can researchers make of this comparative review? It may be tempting to think that the only point in the discussion that really matters is the first one, and that researchers should simply use whichever approach is consistent with the data confronting them. However, researchers who actually plan their analyses before collecting their data will need to consider both alternatives. Very often, at this point researchers have the freedom to design their study to be consistent with either method. Even the ex post analyst needs to understand the trade-offs involved, when difficulties with the natural approach create the temptation to switch to the alternative.

The indicant product approach avoids the sample splitting required by the multisample approach, and may be more parsimonious. On the other hand, the indicant product interaction model is only well developed for the multiplicative case—and a centering transformation limits this approach to "consistency effects." The multisample approach is much more flexible, and it avoids the multicollinearity and distributional problems that can be associated with the indicant product approach. Although recent developments have made it easier to specify models under the indicant product approach, the multisample approach can be implemented using virtually any popular SEM software package. In sum, researchers will still find the multisample approach to be more suitable under the greatest variety of circumstances, especially given discrete or categorical interaction variables.

DEMONSTRATING THE MULTISAMPLE APPROACH

To help researchers apply the multisample approach, we demonstrate fitting the MacKenzie and Spreng (1992) example using two different software packages—Amos 3.5 (Arbuckle, 1995) and LISREL 8 (Jöreskog & Sörbom, 1993). In each case, there are two estimated models. The first allows the means of the A_{ad} and A_{brand} constructs and the value of the β parameter to differ in the low- and high-Motivation samples, whereas the second model constrains those three parameters to equality across groups. (In these examples, all other parameters are constrained to equality across groups.) Thus, the two models are nested, and, given assumptions, the difference in their chi-square statistics will be a chi-square difference with 3 degrees of freedom (*df*). Command files for the software packages are included in the Appendix. The necessary data are reprinted in Table 1.1.

LISREL 8 computes the chi-square statistics for the two models as 41.30 (32 *df*, p = .13) and 50.37 (35 *df*, p = .045). The difference, 9.07 (3 *df*) is significant (p = .028), so the interaction hypothesis is supported. AMOS 3.5 computes the chi-square statistics for the two models as 41.331 (32 *df*, p = .125) and 50.436 (35 *df*, p = .044). These very slight differences may be due to the different optimization methods used by the two programs. The chi-square difference statistics for the combinations of slope and mean constraints are summarized in the margins of Table 1.2. Amos computes the difference chi-square statistics between the interaction and the group-invariant models as 9.105 (df = 3, p = .028). This is essentially the same result obtained with LISREL. The small numeric difference of the fit statistics computed by Amos and LISREL8 arises because AMOS fits the maximum likelihood sample covariance matrices (i.e., with N_g as the denominator) by default, whereas LISREL computes the model fit using bias-corrected sample covariance matrices (i.e., with $N_g - 1$ as the denominator).

The multiple model comparisons summarized in Table 1.2 indicate a difference in slope between the low- and high-Motivation groups, but not a

TABLE 1.1
Data for the MacKenzie and Spreng (1992) Example

Low-Motivation Group								
Variable	μ	σ^2	*Correlations*					
Y_1	4.27	1.23	1					
Y_2	5.02	1.14	0.64	1				
Y_3	4.48	1.26	0.78	0.73	1			
Y_4	4.69	1.19	0.68	0.63	0.69	1		
Y_5	4.53	1.29	0.43	0.55	0.50	0.59	1	
Y_6	4.66	1.28	0.65	0.63	0.67	0.81	0.60	1
High-Motivation Group								
Variable	μ	σ^2	*Correlations*					
Y_1	4.35	1.24	1					
Y_2	4.93	1.18	0.72	1				
Y_3	4.59	1.20	0.76	0.74	1			
Y_4	4.86	1.10	0.51	0.46	0.57	1		
Y_5	4.71	1.08	0.32	0.33	0.39	0.40	1	
Y_6	4.74	1.15	0.54	0.45	0.60	0.73	0.45	1

Note. From "How Does Motivation Moderate the Impact of Central and Peripheral Processing on Brand Attitudes and Intentions" by S. B. MacKenzie and R. A. Spreng, 1992, *Journal of Consumer Research*, *18*, 525. Copyright © 1992 by The University of Chicago Press. Adapted by permission.

TABLE 1.2
Model Fit Statistics for Four Combinations of Slope and Mean Constraints

	Same Means	*Different Means*	*Test of Equal Means*
Same Slopes	$\chi^2 = 50.4$, $df = 35$	$\chi^2 = 48.9$, $df = 33$	$\chi^2 = 1.5$, $df = 2$
Different Slopes	$\chi^2 = 42.9$, $df = 34$	$\chi^2 = 41.3$, $df = 32$	$\chi^2 = 1.6$, $df = 2$
Testing Equality of Slopes	$\chi^2 = 7.5$, $df = 1$	$\chi^2 = 7.6$, $df = 1$	$\chi^2 = 9.1$, $df = 3$

difference in means. These two comparisons represent two specific tests of hypotheses out of a very large number of possibilities. In principle, every parameter of the model could be specified to be equal or different across the groups, and all possible combinations of parameters set equal or different could be tested. Such explorations are not recommended; it is best to limit the analyses to tests of specific hypotheses such as those given in this example.

These examples illustrate the simplicity with which interactions may be modeled with standard SEM software. However, we should end with a final note of caution concerning widespread application of the method. An assumption of the method is that the interaction variable—the one used to assign subjects to one or other group—does not covary with the observed variables in the model. That is, there is no main effect of the interacting variable. Should such a relationship exist, and not be controlled, it could be interpreted as evidence for an interaction where only a linear relationship exists.

REFERENCES

Anderson, J. C., & Gerbing, D. W. (1988). Structural equation modeling in practice: A review and recommended two-step approach. *Psychological Bulletin, 103*(3), 411–423.

Arbuckle, J. (1995). *Amos users' guide*. Chicago: SmallWaters.

Bagozzi, R. P., & Yi, Y. (1989). On the use of structural equation models in experimental designs. *Journal of Marketing Research, 26*(3), 271–284.

Bentler, P. M. (1993). *EQS: Structural equations program manual*. Los Angeles: BMDP.

Bollen, K. A. (1989). *Structural equations with latent variables*. New York: Wiley.

Bollen, K. A., & Stine, R. A. (1993). Bootstrapping goodness-of-fit measures in structural equation models. In K. A. Bollen & J. S. Long (Eds.), *Testing structural equation models* (pp. 111–135). Newbury Park, CA: Sage.

Cliff, N. (1983). Some cautions concerning the application of causal modeling methods. *Multivariate Behavioral Research, 18,* 115–126.

Fraser, C. (1980). *COSAN user's guide*. Toronto: The Ontario Institute for Studies in Education.

Hartmann, W. (1995). *The CALIS procedure: Release 6.11 extended user's guide*. Cary, NC: SAS Institute.

Hayduk, L. A. (1987). *Structural equation modeling with LISREL: Essentials and advances*. Baltimore: Johns Hopkins University Press.

Jaccard, J., & Wan, C. K. (1996). *LISREL approaches to interaction effects in multiple regression*. Thousand Oaks, CA: Sage.

Jöreskog, K. G., & Sörbom, D. (1993). *LISREL 8 user's reference guide*. Chicago: Scientific Software.
Jöreskog, K. G., & Yang, F. (1996). Non-linear structural equation models: The Kenny–Judd model with interaction effects. In G. A. Marcoulides & R. E. Schumacker (Eds.), *Advanced structural equation modeling: Issues and techniques* (pp. 57–88). Mahwah, NJ: Lawrence Erlbaum Associates.
Kenny, D. A., & Judd, C. M. (1984). Estimating the nonlinear and interactive effects of latent variables. *Psychological Bulletin, 96*(1), 201–210.
Keppel, G., & Zedeck, S. (1989). *Data analysis for research designs: Analysis of variance and multiple correlation/regression approaches*. New York: Freeman.
Lee, S.-Y., Poon, W.-Y., & Bentler, P. M. (1992). Structural equation models with continuous and polytomous variables. *Psychometrika, 57*(1), 89–105.
Lilien, G. L., & Kotler, P. (1983). *Marketing decision making: A model-building approach*. New York: Harper & Row.
MacCallum, R. C., Roznowski, M., & Necowitz, L. B. (1992). Model modifications in covariance structure analysis: The problem of capitalization on chance. *Psychological Bulletin, 111*(3), 490–504.
MacKenzie, S. B., & Spreng, R. A. (1992). How does motivation moderate the impact of central and peripheral processing on brand attitudes and intentions. *Journal of Consumer Research, 18*(1), 519–529.
Muthén, B., & Kaplan, D. (1992). A comparison of some methodologies for the factor analysis of non-normal Likert variables: A note on the size of the model. *British Journal of Mathematical and Statistical Psychology, 45,* 19–30.
Neale, M. C. (1995). *Mx: Statistical modeling*. Box 710 MCV, Richmond, VA 23298: Department of Psychiatry. 3rd edition.
Ollson, U. (1979). On the robustness of factor analysis against crude classification of the observations. *Multivariate Behavioral Research, 14,* 485–500.
Ping, R. A. (1995). A parsimonious estimating technique for interaction and quadratic latent variables. *Journal of Marketing Research, 32,* 336–347.
Rigdon, E. E., & Ferguson, C. E., Jr. (1991). The performance of the polychoric correlation coefficient and selected fitting functions in confirmatory factor analysis with ordinal data. *Journal of Marketing Research, 28*(4), 491–497.
Rindskopf, D. (1984). Using phantom and imaginary latent variables to parameterize constraints in linear structural models. *Psychometrika, 49*(1), 37–47.
Russell, C. J., & Bobko, P. (1992). Moderated regression analysis and likert scales: Too coarse for comfort. *Journal of Applied Psychology, 77*(2), 336–342.
Smith, K. W., & Sasaki, M. S. (1979). Decreasing multicollinearity: A method for models with multiplicative functions. *Sociological Methods & Research, 8*(3), 35–56.
Wothke, W. (1993). Nonpositive definite matrices in structural modeling. In K. A. Bollen & J. S. Long (Eds.), *Testing structural equation models* (pp. 256–293). Newbury Park, CA: Sage.
Yung, Y.-F., & Bentler, P. M. (1994). Bootstrap-corrected ADF test statistics in covariance structure analysis. *British Journal of Mathematical and Statistical Psychology, 47*(2), 63–84.

APPENDIX

model

```
KM
SD
ME
LA
ATT_AD1 ATT_AD2 ATT_AD3 BRANDAT1 BRANDAT2 BRANDAT3
MO NE=2 NY=6 LY=FU,FI BE=FU,FI PS=DI TE=DI TY=FR AL=FI
LE
ATT_AD ATT_BRND
PA LY
0 0
2(1 0)
0 0
2(0 1)
VA 1 LY(1,1) LY(4,2)
FR BE(2,1)
OU
HIGH MOTIVATION SAMPLE
DA NO=160
KM
SD
ME
LA
ATT_AD1 ATT_AD2 ATT_AD3 BRANDAT1 BRANDAT2 BRANDAT3
MO LY=IN BE=PS TE=IN PS=IN TY=IN AL=FR
LE
ATT_AD ATT_BRND
OU PT
```

LISREL command file for the constrained model:

```
LOW MOTIVATION SAMPLE
DA NI=6 NO=200 MA=CM NG=2
KM
SD
ME
LA
ATT_AD1 ATT_AD2 ATT_AD3 BRANDAT1 BRANDAT2 BRANDAT3
MO NE=2 NY=6 LY=FU,FI BE=FU,FI PS=DI TE=DI TY=FR AL=FI
LE
ATT_AD ATT_BRND
PA LY
0 0
2(1 0)
0 0
2(0 1)
VA 1 LY(1,1) LY(4,2)
FR BE(2,1)
OU
```

```
HIGH MOTIVATION SAMPLE
DA NO=160
KM
SD
ME
LA
ATT_AD1 ATT_AD2 ATT_AD3 BRANDAT1 BRANDAT2 BRANDAT3
MO LY=IN BE=IN TE=IN PS=IN TY=IN AL=IN
LE
ATT_AD ATT_BRND
OU PT
```

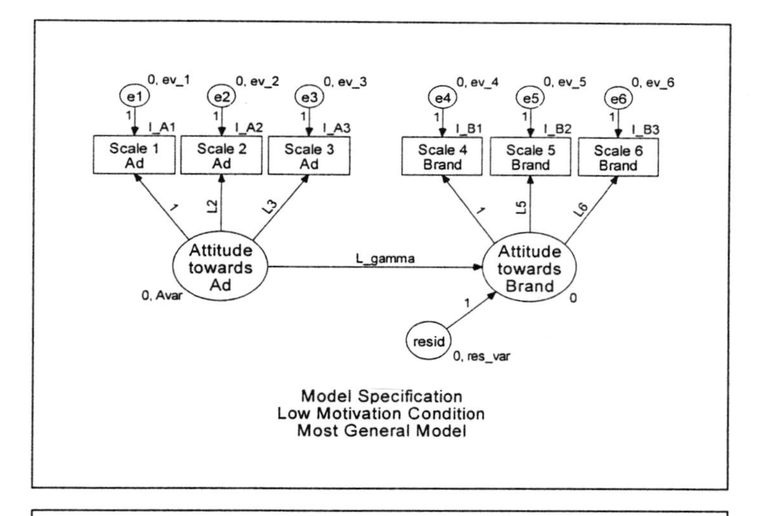

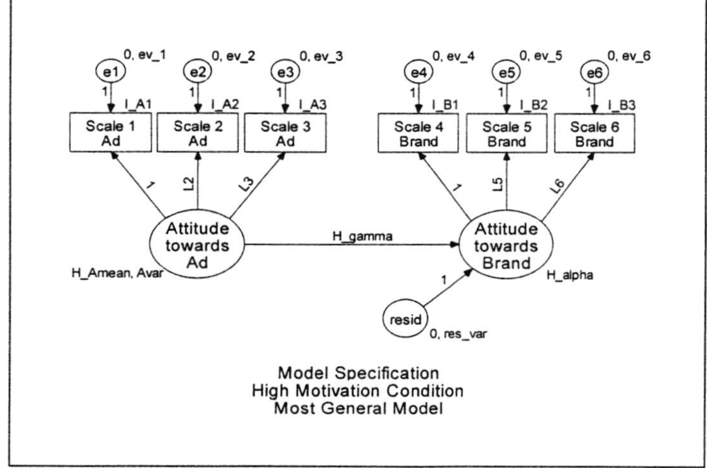

FIG. 1.2. Two-Group Interaction Model, Specification with Amos Graphics.

Command File for Interaction Model Tests with Amos

```
!Model constraints for multiple model comparisons with Amos

$Model = Most_General_Model ! No constraints

$Model = Same_slopes          ! Same gamma coefficients
   L_gamma = H_gamma          ! in both groups

$Model = Same_means           ! Same group means for
   H_Amean = 0                ! attitudes to Ad as well
   H_alpha = 0                ! as to Brand

$Model = Same_regressions     ! This model combines both
   Same_means                 ! sets of constraints for
   Same_slopes                ! group means and slopes
```

2

Modeling Interaction and Nonlinear Effects: A Step-by-Step LISREL Example

Fan Yang Jonsson
Uppsala University, Uppsala, Sweden

An example of a theory involving interactions among theoretical constructs is the theory of reasoned action or expectancy-value attitude theory in social psychology (Ajzen & Fishbein, 1980; Fishbein & Ajzen, 1975). See Bagozzi (1986) for further explanation of this theory and Bagozzi, Baumgartner, and Yi (1992) and references therein for its interpretation and application in consumer research. In essence, this theory states that beliefs about consequences of an act and evaluations of consequences of the same act interact in determining a person's overall attitude or intention to perform the act. This theory can be and has been applied to many different acts, such as using public transportation rather than a private car to commute to work, or donating or not donating blood or an organ. There are various extensions of this theory, but for the present illustration, only the basic theory referred to here is used.

In this chapter various approaches for testing this theory and estimating the interaction effect are illustrated using a small data set.

VARIABLES

In Bagozzi et al. (1992), the attitudinal act of interest was consumers' use of coupons for grocery shopping, and the two consequences of coupon use were the beliefs that using coupons would lead to savings on the grocery bill and using coupons would contribute to a feeling of being a good shopper.

Beliefs about consequences of using coupons were measured by the responses to the following:

> Please indicate how likely it is that each of the following consequences would occur to you if you were to use coupons for shopping in the supermarket in the upcoming week. (Ignore how important each consequence is to you at this time. Here we would like you to simply indicate the likely amount of each consequence occurring to you if you were to use coupons.)

BE1: Much time and effort would be required to search for, gather, and organize coupons.

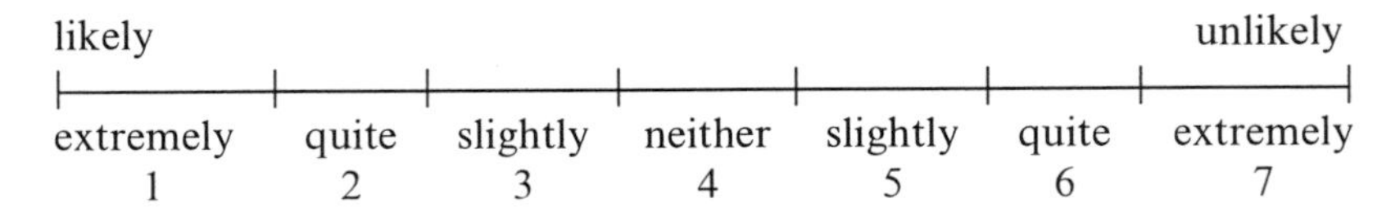

BE2: Much time and effort would be required to plan the use of and actually redeem coupons in the supermarket.

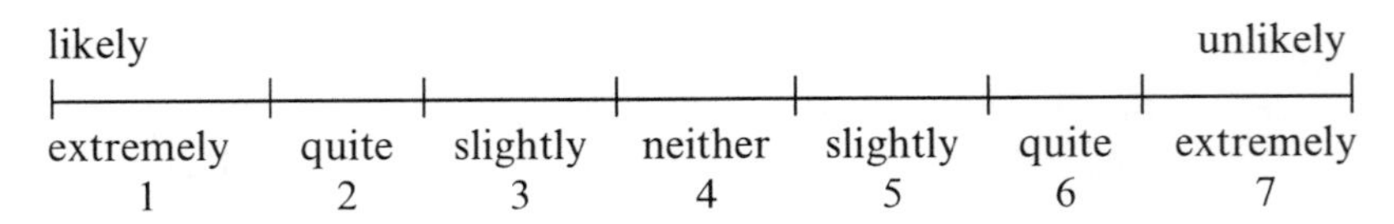

Evaluations of consequences of using coupons were measured by responses to the following:

> Please express your evaluations of each of the following consequences of using coupons for shopping in the supermarket in the upcoming week.

VL1: The time and effort required to search for, gather, and organize coupons makes me feel:

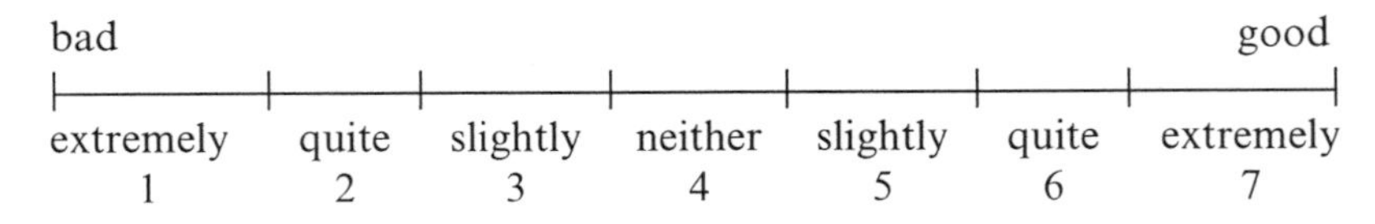

VL2: The time and effort required to plan the use of and actually redeem coupons in the supermarket makes me feel:

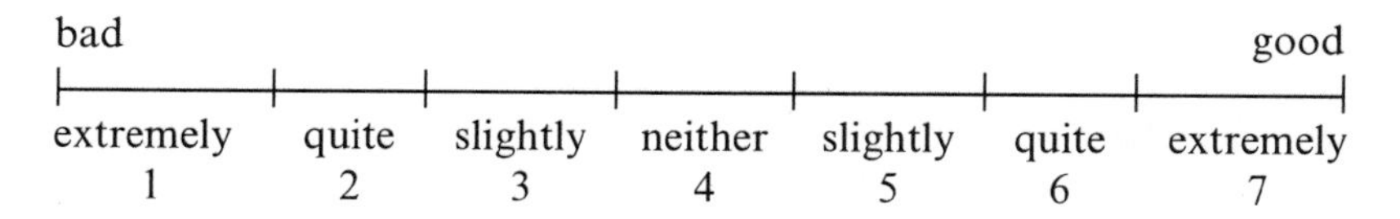

The overall attitude toward coupons was measured by the responses to the following:

My attitude toward coupons can be best summarized as

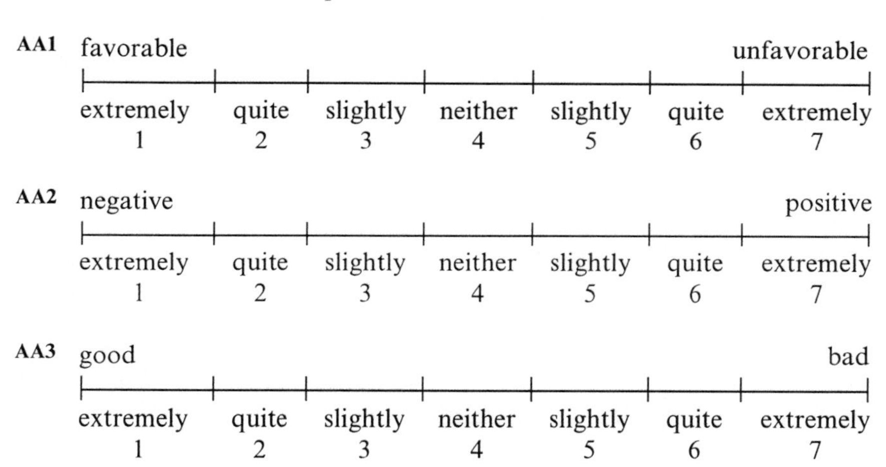

After data collection, items **AA1** and **AA3** were recoded in the opposite direction.

DATA AND MODEL

The data[1] come from a survey of 253 female staff members at two large public universities in the United States. The term *BEA data* is used as a working name for this data set (BEA = Beliefs, Evaluations, Attitudes).

Writing

$$\eta = \mathbf{At} = \textit{attitude},$$
$$\xi_1 = \mathbf{Be} = \textit{belief},$$
$$\xi_2 = \mathbf{Vl} = \textit{evaluation},$$

the expectancy-value model is

[1]The author is grateful to Professor Richard P. Bagozzi for kindly providing these data.

$$\mathbf{At} = \alpha + \gamma_1\,\mathbf{Be} + \gamma_2\,\mathbf{Vl} + \gamma_3(\mathbf{Be} \times \mathbf{Vl}) + \zeta,$$

or in LISREL notation

$$\eta = \alpha + \gamma_1\xi_1 + \gamma_2\xi_2 + \gamma_3\xi_1\xi_2 + \zeta. \tag{1}$$

There are three indicators **AA1**, **AA2**, **AA3** of **At**, two indicators **BE1** and **BE2** of **Be**, and two indicators **VL1** and **VL2** of **Vl**. In LISREL notation these correspond to y_1, y_2, y_3, x_1, x_2, x_3, x_4, respectively. The model is similar to the Kenny–Judd model (Kenny & Judd, 1984) and differs from it only in that there are three indicators of η rather than only one.

When there are two indicators of each independent construct, one has a choice of using one or more product variables. Kenny and Judd (1984) suggested the use of four product variables. This seems like a "natural" choice because there are four natural indicators of $\xi_1\xi_2$, namely x_1x_3, x_1x_4, x_2x_3, and x_2x_4. For linear factor analysis it is a generally accepted principle to use as many indicators as possible. The more that are used the better the measurement of the latent variable will be. This suggests that four product variables should be used as indicators of the latent product variable $\xi_1\xi_2$. However, as shown by Jöreskog and Yang (1996), only one product variable is necessary to identify the model. So, in principle, it is possible to choose one, two, three, or four product variables. Kenny and Judd did not discuss this choice, they simply took it for granted that four product variables should be used.

Is it better to use four product variables or only one? As stated previously, using four product variables is not necessary because the model is identified with just one product variable. Three more product variables adds many more manifest parameters without adding any new parameters to estimate. So it would seem that one gains many degrees of freedom and that the model therefore is much more parsimonious. However, there is a price paid for this. First, the augmented moment matrix **A** necessary for a correct analysis (see the section, Full Information Estimation) will be larger when more product variables are used, and the corresponding estimated asymptotic covariance matrix $\mathbf{W}_a$ will be more rank deficient (see Jöreskog & Yang, 1996).

The choice of one versus four product variables was studied by Yang Jonsson (1997) using simulation methods. It was found that four product variables gave less bias, but it underestimated standard errors more than one product variable.

When **Be** and **Vl** are measured using questions with the same issues, as is the case here, it seems natural to use only product variables corresponding to the same issues as indicators of the latent product variable. This means that only two product variables should be used, namely **BE1VL1** = x_1x_3 and

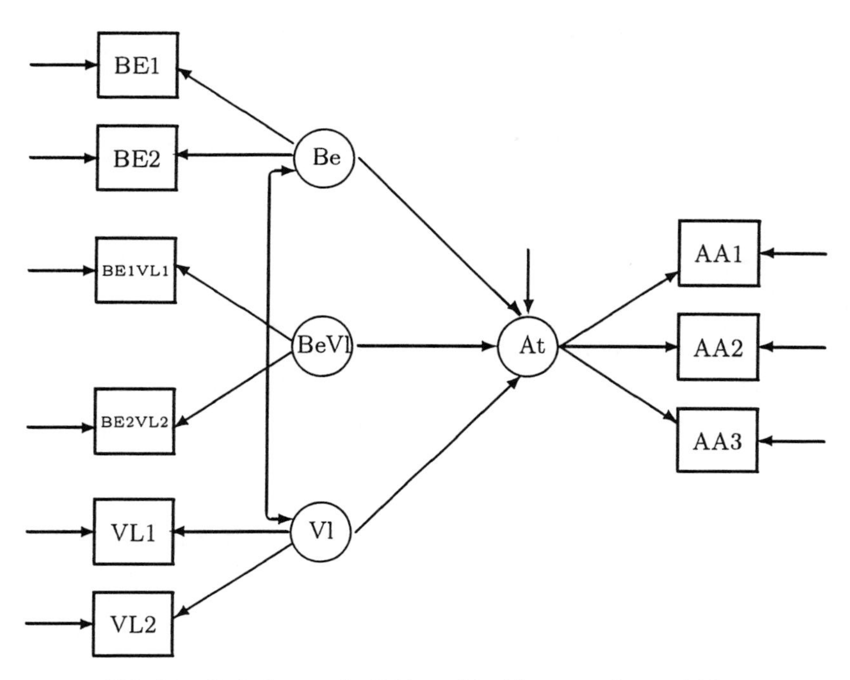

FIG. 2.1. Path diagram for BEA model with two product variables.

BE2VL2 = x_2x_4. A path diagram for the model with two product variables is shown in Fig. 2.1.

It is clear from the questionnaire items that **Be** should have a positive effect on **At** because a person responding in the low end of the scale on **BE1** and **BE2** is also likely to respond in the low end on each of **AA1**, **AA2**, and **AA3**; and a person responding in the high end of the scale on **BE1** and **BE2** is also likely to respond in the high end on each of **AA1**, **AA2**, and **AA3**. Similarly, it is clear that there should be a positive effect of **Vl** on **At**.

If the regression coefficient of **Be** on **At** is denoted b_1 and that of **Vl** on **At** is denoted b_2, the theory of reasoned action predicts that there is an interaction effect of **Be** and **Vl** on **At**, such that the regression coefficient of **Be** on **At** is smaller than b_1 for those below the mean of **Vl** and larger than b_1 for those above the mean of **Vl**. In the same way, the theory predicts that the regression coefficient of **Vl** on **At** is smaller than b_2 for those below the mean of **Be** and larger than b_2 for those above the mean of **Be**. This idea can be used to check whether there is an interaction effect.

It should be noted, however, that the theory does not state that there must be main effects of **Be** and **Vl** on **At**, only that there is an interaction effect. It is possible that either γ_1 or γ_2 or both are zero, but γ_3 must be positive.

The data set has the following disadvantages:

1. The variables are ordinal. However, because the model requires data on interval scales, the ordinality will be ignored and the scores 1 through 7 will be treated as if they have metric properties.

2. Because there are only a few indicators of each theoretical construct (latent variable), there is a very low construct validity. It would have been better if the investigators had asked several more questions tapping each construct. Not only would this increase construct validity; it would also make it possible to construct more continuous scales. For example, if there had been 10 questions on **Be**, one could use the sum of the first 5 and the last 5 as two indicators **BE1** and **BE2** of **Be**, and each of these would be more continuous and more normally distributed.

3. The sample size was only 253. In view of the results on sample size reported by Yang Jonsson (1997), this may be too small to estimate the model reasonably and there may be too little power to detect an interaction effect, if one exists.

A PRELIS command file for computing the sample mean vector $\bar{\mathbf{z}}$, the sample covariance matrix **S**, and the asymptotic covariance matrix **W** needed for maximum likelihood (ML) and weighted least squares (WLS) are given in Appendix A. A PRELIS input file for computing the augmented moment matrix **A** and the corresponding asymptotic covariance matrix $\mathbf{W}_a$ is also given.

The means, standard deviations, and intercorrelations of the variables, including the two product variables, are given in Table 2.1. For the originally observed variables, the correlations are higher within each set of indicators than between sets. This is consistent with the validity of the measurement model for the three construct variables **At**, **Be**, and **Vl**. This is tested later.

If (ξ_1, ξ_2) and (δ_1, δ_2, δ_3, δ_4) are independent and normally distributed, and the model holds, the observed variables x_1, x_2, x_3, x_4 would be normally

TABLE 2.1
Means, Standard Deviations, and Correlations of Observed Variables

Variables	*Correlations*								
AA1	1.00								
AA2	0.67	1.00							
AA3	0.68	0.74	1.00						
BE1	0.27	0.39	0.38	1.00					
BE2	0.23	0.34	0.31	0.60	1.00				
VL1	0.50	0.55	0.58	0.51	0.47	1.00			
VL2	0.37	0.47	0.41	0.39	0.53	0.65	1.00		
BE1VL1	0.41	0.52	0.53	0.92	0.62	0.80	0.57	1.00	
BE2VL2	0.35	0.47	0.43	0.60	0.91	0.64	0.81	0.71	1.00
Means	4.51	5.04	5.00	5.12	5.34	5.81	5.72	30.35	31.23
St. Dev.	1.36	1.31	1.42	1.33	1.36	0.91	0.93	10.77	10.92

distributed. However, as already stated, the variables are ordinal and therefore not normally distributed. Even if they are treated as metric variables, a simple data screening with PRELIS reveals that they are non-normal. All the omnibus tests of normality of **BE1, BE2, VL1, VL2** are significant at the 5% level. Hence, it is not likely that these assumptions hold. This suggests that WLS or WLSA (weighted least squares based on the augmented moment matrix) would be more appropriate to use than ML.

PRELIMINARY ANALYSIS

Before estimating the general model specified in the previous section, the measurement model for **At**, **Be**, and **Vl** is tested and a check of the evidence for an interaction between **Be** and **Vl** is also made. If the measurement model is not valid, the whole nonlinear model becomes invalid. If the measurement model is valid but there is no evidence of an interaction effect, it is best to use a linear model.

Subgroup Analysis

Two subgroups are formed by splitting at the mean of **BE** = **BE1** + **BE2**. The mean is 10.454. This gives 110 cases in the low group and 143 cases in the high group. The existence of an interaction effect can be checked by a multigroup analysis of the model shown in Fig. 2.2.

Assuming that the factor loadings are invariant over groups, the model with unequal γs gives $\hat{\gamma}_{low} = -0.16(0.29)$ and $\hat{\gamma}_{high} = 1.26(0.45)$ and an overall fit of $\chi^2 = 19.01$ with 11 degrees of freedom. The model with equal γs gives $\hat{\gamma} = 1.24(0.45)$ and an overall fit of $\chi^2 = 22.99$ with 12 degrees of freedom. The difference in chi-square, 3.98, can be used as test of the hypothesis that the two γs are equal. This is significant at the 5% level, thus giving support for an interaction effect. Although the difference between $\hat{\gamma}_{high}$ and $\hat{\gamma}_{low}$ is considerable, $\hat{\gamma}_{low}$ is not significant from zero. So the regression in the low group is mostly "noise." This explains why $\hat{\gamma}$ is so close to $\hat{\gamma}_{high}$ and why the chi-square difference is small, but significant.

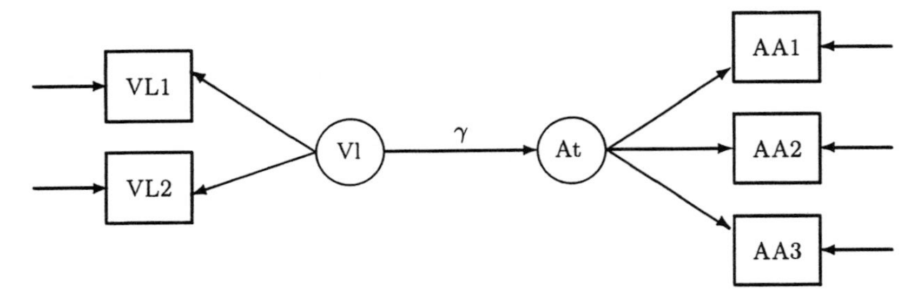

FIG. 2.2. Path diagram for subgroup regression of **At** on **Vl**.

One can also form subgroups by splitting at the mean of **VL** = **VL1** + **VL2**. The mean is 11.538. This gives 114 cases in the low group and 139 cases in the high group. Assuming again that the factor loadings are invariant over groups, the model with unequal γs gives $\hat{\gamma}_{low}$ = 0.05(0.13) and $\hat{\gamma}_{high}$ = 0.58(0.17) and an overall fit of χ^2 = 12.27 with 11 degrees of freedom. The model with equal γs gives $\hat{\gamma}$ = 0.41(0.10) and an overall fit of χ^2 = 19.68 with 12 degrees of freedom. The difference in chi-square, 7.41, is significant at the 1% level, thus giving more support for an interaction effect. Input files for this approach are given in Appendix B.

This approach, although useful as a preliminary check of the evidence for an interaction effect, has certain disadvantages:

1. The grouping is ad hoc and can lead to different results. It also reduces power.
2. The grouping is based on an observed variable and not on the latent variable, and it ignores measurement error in that observed variable.
3. It does not give an estimate of γ_3.

Testing the Measurement Model

The measurement model can be tested by specifying a confirmatory factor analysis model with three correlated factors as shown in Fig. 2.3.

Because the observed variables are non-normal, the model is fitted using WLS estimation. This results in estimates of factor loadings, measurement error variances, and factor covariances shown in Table 2.2.

The overall fit of the model is given by χ^2 = 15.73 with 11 degrees of freedom. This has a p value of .15, which shows that the measurement model is valid. Appendix C gives input files for testing the measurement model.

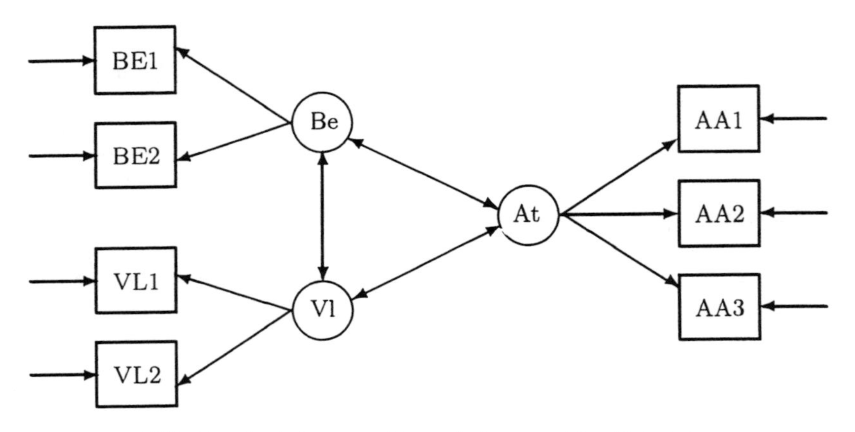

FIG. 2.3. Path diagram of three-factor model for BEA data.

TABLE 2.2
Parameter Estimates for Measurement Model

	Factor Loadings			
Measure	*At*	*Be*	*Vl*	*Error Variance*
AA1	1.00	0	0	0.69
AA2	1.05	0	0	0.47
AA3	1.15	0	0	0.46
BE1	0	1.00	0	0.73
BE2	0	1.14	0	0.33
VL1	0	0	1.00	0.13
VL2	0	0	0.85	0.32

	Factor Covariance Matrix		
	At	*Be*	*Vl*
At	1.16		
Be	0.53	0.84	
Vl	0.63	0.56	0.69

Estimating the Interaction Effect Using Factor Scores

The factor covariance matrix in Table 2.2 is a good estimate of the covariance matrix of the three construct variables **At**, **Be**, and **Vl**, in which the effects of measurement errors in the observed variables have been eliminated. However, in order to estimate Equation 1, this must be extended with the latent product variable **BeVl**. The problem is that there is no way of doing this without making strong assumptions.

One approach to solving this problem is to estimate factor scores for **At**, **Be**, and **Vl** for each individual in the sample, and then compute the product of the factor scores for **Be** and **Vl**. After that, one can proceed as if **At**, **Be**, **Vl**, and **BeVl** are observed.

This approach has certain drawbacks. First, there are no factor scores available that satisfy all the properties of the factors of which there are estimates (see, e.g., Lawley & Maxwell, 1971). Second, because this is a two-step procedure, there is no guarantee that the standard errors obtained in the second step are reasonably correct. Nevertheless, to get a first estimate of the γs in Equation 1 this approach is worth pursuing.

To do so, factor scores derived by Anderson and Rubin (1956, Equations 9.6 and 9.8) are used. These are unbiased and can be estimated subject to the constraint that their sample covariance matrix equals the estimated factor covariance matrix exactly.

Using observed scores in deviation from their means and denoting the estimated factor scores for **At**, **Be**, and **Vl**, by **AT**, **BE**, and **VL**, respectively, these factor score estimates are computed for each case in the sample as

$$\begin{pmatrix} \mathbf{AT} \\ \mathbf{BE} \\ \mathbf{VL} \end{pmatrix} = \begin{pmatrix} 0.195 & 0.303 & 0.333 & 0.009 & 0.023 & 0.053 & 0.018 \\ 0.010 & 0.015 & 0.017 & 0.188 & 0.474 & 0.110 & 0.037 \\ 0.010 & 0.015 & 0.017 & 0.019 & 0.048 & 0.656 & 0.222 \end{pmatrix} \begin{pmatrix} \mathbf{AA1} \\ \mathbf{AA2} \\ \mathbf{AA3} \\ \mathbf{BE1} \\ \mathbf{BE2} \\ \mathbf{VL1} \\ \mathbf{VL2} \end{pmatrix}$$

where the matrix is computed from Equation 9.6 in Anderson and Rubin. This results in a data matrix of order 253 × 3. A fourth column, **BEVL** = **BE** × **VL**, is added. The covariance matrix of these estimated factor scores, including the product variable, is given in Table 2.3. The means of the first three factor scores are zero, and the mean of the product variable is 0.561. The PRELIS input file for computing this covariance matrix is given in Appendix D.

Using this covariance matrix, one can easily estimate a number of different regression equations with **AT** as a dependent variable. Various regression estimates are given in Table 2.4. Each regression equation has an estimated intercept, which is of no interest here, and is therefore not reported. Each entry in the table gives the regression estimate followed by the standard error in parenthesis followed by the *t* value.

The linear model 4 has $R^2 = 0.50$ but **BE** is not significant. This is almost the same as Equation 2, where only **VL** is included. In Equation 5, both **BE** and **BEVL** are significant but R^2 is only 0.30. Equation 6 has both **VL** and **BEVL** but only **VL** is significant, so this is again essentially the same as 2. With all three variables included, as in Equation 7, again only **VL** is significant, so this again leads to Equation 2.

If one takes these results at face value, one must say that the evidence for an interaction effect is very weak. In fact, only **VL** seems to have an effect on **AT**. However, there are reasons to believe that these results might be misleading. The estimated factor scores for each person are based on the estimates of factor loadings, factor variances and covariances, and error variances in Table 2.2, which in turn are estimated from the data for all

TABLE 2.3
Covariance Matrix of Factor Scores

	Factor Covariance Matrix			
	AT	*BE*	*VL*	*BEVL*
AT	1.157			
BE	0.527	0.844		
VL	0.628	0.561	0.689	
BEVL	0.025	−0.133	−0.027	0.608

TABLE 2.4
Regression Estimates

	BE	*VL*	*BEVL*	R^2
1	0.620(0.066)9.52	—	—	0.28
2	—	0.910(0.062)14.8	—	0.49
3	—		0.040(0.100)0.39	0.00
4	0.039(0.078)0.50	0.880(0.093)9.42	—	0.50
5	0.650(0.067)9.70	—	0.180(0.076)2.40	0.30
6	—	0.910(0.060)15.2	0.081(0.062)1.32	0.50
7	0.066(0.077)0.85	0.860(0.094)9.13	0.093(0.060)1.55	0.50

persons. This introduces dependence between persons. Thus, the data on the factor scores cannot be regarded as a sample of independent observations, which is a crucial assumption in the formulae for the standard errors.

FULL INFORMATION ESTIMATION

Although the approaches discussed in the preceding sections may be useful as preliminary steps, the only way to estimate the model correctly is to specify all the constraints correctly and estimate all free parameters of the model simultaneously. As shown by Jöreskog and Yang (1996), both the mean vector and covariance matrix are constrained. So, both the sample mean vector and the sample covariance matrix must be used to fit the model.

Let $\bar{\mathbf{z}}$ and $\mathbf{S}$ be the sample mean vector and covariance matrix. Let μ and Σ be the corresponding population mean vector and covariance matrix. These are functions of the parameter vector θ containing all the independent parameters in the model. To estimate the model, $\mu(\theta)$ and $\Sigma(\theta)$ are fitted to $\bar{\mathbf{z}}$ and $\mathbf{S}$ simultaneously. Three methods, ML, WLS, and WLSA are used for estimation. In particular, WLSA should give correct asymptotic standard errors and chi-square goodness-of-fit measures.

Maximum Likelihood

The ML approach estimates θ by minimizing the fit function

$$F(\theta) = \log\|\Sigma\| + \mathrm{tr}(\mathbf{S}\Sigma^{-1}) - \log\|\mathbf{S}\| - k + (\bar{\mathbf{z}} - \mu)'\Sigma^{-1}(\bar{\mathbf{z}} - \mu). \qquad (2)$$

where k is the number of variables in $\mathbf{z}$ (here $k = 6$). This fit function is derived from the ML principle based on the assumption that the observed variables $\mathbf{z}$ have a multinormal distribution (see, e.g., Anderson, 1984, for a derivation of the multinormal likelihood). Because of the product variables, this assumption does not hold for the model considered here. As stated previously, the random vector $\mathbf{z}$ is not multivariate normal. Nevertheless, the ML approach is worth consideration because it may be robust against the kind of non-normality generated by the Kenny–Judd (1984) model.

Weighted Least Squares

The WLS approach estimates θ by minimizing the fit function

$$F(\theta) = (\mathrm{s} - \sigma)'\mathbf{W}^{-1}(\mathrm{s} - \sigma) + (\bar{\mathbf{z}} - \mu)'\mathbf{S}^{-1}(\bar{\mathbf{z}} - \mu), \tag{3}$$

where

$$\mathrm{s}' = (s_{11}, s_{21}, s_{22}, s_{31}, \ldots, s_{kk}),$$

is a vector of the nonduplicated elements $\mathbf{S}$,

$$\sigma' = (\sigma_{11}, \sigma_{21}, \sigma_{22}, \sigma_{31}, \ldots, \sigma_{kk}),$$

is the vector of corresponding elements of $\Sigma(\theta)$ reproduced from the model parameters θ, and $\mathbf{W}$ is a symmetric positive definite matrix. The usual way of choosing $\mathbf{W}$ in WLS is to let $\mathbf{W}$ be a consistent estimate of the asymptotic covariance matrix of s. In the general case when the observed variables are continuous and have a multivariate distribution satisfying very mild assumptions, the covariance matrix of s has a typical element (see Browne, 1984)

$$\omega_{gh,ij} = Cov(s_{gh}, s_{ij}) = \mu_{ghij} - \sigma_{gh}\sigma_{ij}, \tag{4}$$

where μ_{ghij} is a fourth-order central moment. This can be estimated as

$$\omega_{gh,ij} = m_{ghij} - s_{gh}s_{ij}, \tag{5}$$

where

$$m_{ghij} = (1/N)\sum_{a=1}^{N}(z_{ag} - \bar{z}_g)(z_{ah} - \bar{z}_h)(z_{ai} - \bar{z}_i)(z_{aj} - \bar{z}_j) \tag{6}$$

is a fourth-order central sample moment. Using such a $\mathbf{W}$ in Equation 3 gives what Browne called “asymptotically distribution free best GLS estimators” (p. 71) for which correct asymptotic chi-squares and standard errors may be obtained under non-normality.

WLS does not assume multivariate normality of $\mathbf{z}$ but it does assume that $\bar{\mathbf{z}}$ and $\mathbf{S}$ are asymptotically independent, an assumption that may not hold in this case. Another disadvantage with WLS is that the matrix $\mathbf{W}$ must be estimated from the data and is therefore subject to sampling fluctuations, which may affect the parameter estimates. To avoid severe effects of this kind, **W** must be estimated from a large sample.

WLS Based on the Augmented Moment Matrix

One way to avoid the problems associated with ML and WLS, is to use the augmented moment matrix

$$\mathbf{A} = (1/N)\sum_{c=1}^{N}\begin{pmatrix}\mathbf{z}_c \\ 1\end{pmatrix}(\mathbf{z}'_c\ 1) = \begin{pmatrix}\mathbf{S}+\bar{\mathbf{z}}\bar{\mathbf{z}}' & \\ \bar{\mathbf{z}}' & 1\end{pmatrix}. \tag{7}$$

This is the matrix of sample moments about zero for the vector **z** augmented with a variable that is constant equal to one for every case. The corresponding population matrix is

$$(\alpha_{ij}) = E\begin{pmatrix}\mathbf{z} \\ 1\end{pmatrix}(\mathbf{z}'\ 1) = \begin{pmatrix}\Sigma+\mu\mu' & \\ \mu' & 1\end{pmatrix}. \tag{8}$$

Note that the last element in these matrices is a fixed constant equal to one.

Let

$$\mathbf{a}' = (a_{11}, a_{21}, a_{22}, a_{31}, \ldots, a_{k+1,k}, 1),$$

be a vector of the nonduplicated elements of **A**, and let

$$\alpha' = (\alpha_{11}, \alpha_{21}, \alpha_{22}, \alpha_{31}, \ldots, \alpha_{k+1,k}, 1),$$

be a vector of the corresponding population moments. We can then form another weighted least squares fit function (WLSA):

$$F(\theta) = (\mathbf{a}-\alpha)'\mathbf{W}_a^{-}(\mathbf{a}-\alpha), \tag{9}$$

where $\mathbf{W}_a$ is a consistent estimate of the covariance matrix of **a** and $\mathbf{W}_a^{-}$ is a Moore–Penrose generalized inverse of $\mathbf{W}_a$. See, for example, Graybill (1969) for a definition of generalized inverse and Satorra (1989) and Koning, Neudecker, and Wansbeek (1993) for its use in covariance structure analysis.

The covariance matrix of **a** has a typical element

$$\omega_{gh,ij} = Cov(a_{gh}, a_{ij}) = \nu_{ghij} - \alpha_{gh}\alpha_{ij}, \tag{10}$$

where ν_{ghij} is a fourth-order moment about zero. This can be estimated as

$$\omega_{gh,ij} = n_{ghij} - a_{gh}a_{ij}, \tag{11}$$

where

$$n_{ghij} = (1/N)\sum_{a=1}^{N} z_{ag}z_{ah}z_{ai}z_{aj} \tag{12}$$

is a fourth-order sample moment about zero. Note that because the last element in **a** is a fixed constant, the last row of $\mathbf{W}_a$ is zero. Hence, $\mathbf{W}_a$ is singular. Its rank is even further reduced, however, as shown in Jöreskog and Yang (1996). This is the reason for using a generalized inverse in Equation 9.

To estimate the model, the mean vector $\bar{\mathbf{z}}$, the covariance matrix **S**, the augmented moment matrix **A**, and the asymptotic covariance matrices **W** and $\mathbf{W}_a$ are computed.

With two product variables, the estimated asymptotic covariance matrix **W** is nonsingular and of order 45 × 45. The estimated asymptotic covariance matrix $\mathbf{W}_a$ is of order 55 × 55 but of rank 52. So $\mathbf{W}_{\bar{a}}$ is of order 55 × 55 and of rank 52. Rank deficiency and its implications were discussed by Jöreskog and Yang (1996).

LISREL Implementation for ML and WLS

For the model with two product variables, the LISREL specification for ML and WLS is

$$\begin{pmatrix} y_1 \\ y_2 \\ y_3 \end{pmatrix} = \begin{pmatrix} \tau_1^{(y)} \\ \tau_2^{(y)} \\ \tau_3^{(y)} \end{pmatrix} + \begin{pmatrix} 1 \\ \lambda_2^{(y)} \\ \lambda_3^{(y)} \end{pmatrix} \eta + \begin{pmatrix} \varepsilon_1 \\ \varepsilon_2 \\ \varepsilon_3 \end{pmatrix}, \tag{13}$$

$$\mathbf{y} = \tau_y + \Lambda_y \eta + \varepsilon,$$

$$\begin{pmatrix} x_1 \\ x_2 \\ x_3 \\ x_4 \\ x_1x_3 \\ x_2x_4 \end{pmatrix} = \begin{pmatrix} \tau_1 \\ \tau_2 \\ \tau_3 \\ \tau_4 \\ \tau_1\tau_3 \\ \tau_2\tau_4 \end{pmatrix} + \begin{pmatrix} 1 & 0 & 0 \\ \lambda_2 & 0 & 0 \\ 0 & 1 & 0 \\ 0 & \lambda_4 & 0 \\ \tau_3 & \tau_1 & 1 \\ \tau_4\lambda_2 & \tau_2\lambda_4 & \lambda_2\lambda_4 \end{pmatrix} \begin{pmatrix} \xi_1 \\ \xi_2 \\ \xi_1\xi_2 \end{pmatrix} + \begin{pmatrix} \delta_1 \\ \delta_2 \\ \delta_3 \\ \delta_4 \\ \delta_5 \\ \delta_6 \end{pmatrix},$$

$$\mathbf{x} = \tau_x + \Lambda_x \xi + \delta,$$

$$\eta = \alpha + \gamma_1\xi_1 + \gamma_2\xi_2 + \gamma_3\xi_1\xi_2 + \zeta, \tag{14}$$
$$\eta = \alpha + B\eta + \Gamma\xi + \zeta.$$

The parameter α in Equation 14 and the three τs in Equation 13 are not identified. One can add a constant to the three τs and subtract the same constant from α without affecting the mean vector of $\mathbf{y}$. Hence, we may set $\alpha = 0$ in Equation 14 and estimate the τs in Equation 13. The estimates of the τs *will not* be equal to the means of the y-variables.

The parameter matrices κ, $\mathbf{\Phi}$, Θ_ε, and Θ_δ are

$$\kappa = \begin{pmatrix} 0 \\ 0 \\ \phi_{21} \end{pmatrix} \quad \mathbf{\Phi} = \begin{pmatrix} \phi_{11} & & \\ \phi_{21} & \phi_{22} & \\ 0 & 0 & \phi_{11}\phi_{22} + \phi_{21}^2 \end{pmatrix}.$$

$$\Theta_\varepsilon = diag(\theta_1^{(\varepsilon)}, \theta_2^{(\varepsilon)}, \theta_3^{(\varepsilon)}),$$

$$\Theta_\delta = \begin{pmatrix} \theta_1 & & & & & \\ 0 & \theta_2 & & & & \\ 0 & 0 & \theta_3 & & & \\ 0 & 0 & 0 & \theta_4 & & \\ \tau_3\theta_1 & 0 & \tau_1\theta_3 & 0 & \theta_5 & \\ 0 & \tau_4\theta_2 & 0 & \tau_2\theta_4 & 0 & \theta_6 \end{pmatrix},$$

where θ_5 and θ_6 are

$$\theta_5 = \tau_1^2\theta_3 + \tau_3^2\theta_1 + \phi_{11}\theta_3 + \phi_{22}\theta_1 + \theta_1\theta_3,$$
$$\theta_6 = \tau_2^2\theta_4 + \tau_4^2\theta_2 + \lambda_2^2\phi_{11}\theta_4 + \lambda_4^2\phi_{22}\theta_2 + \theta_2\theta_4.$$

A `LISREL` command file is given in Appendix E.

LISREL Implementation for WLSA

For WLSA, one must treat all observed variables as y-variables and all latent variables as η-variables. This includes the variable that is constant equal to one for every case, which is the only x-variable in the model. The `LISREL` specification is

$$\begin{pmatrix} y_1 \\ y_2 \\ y_3 \\ x_1 \\ x_2 \\ x_3 \\ x_4 \\ x_1x_3 \\ x_2x_4 \end{pmatrix} = \begin{pmatrix} 1 & 0 & 0 & 0 & \tau_1^{(y)} \\ \lambda_2^{(y)} & 0 & 0 & 0 & \tau_2^{(y)} \\ \lambda_3^{(y)} & 0 & 0 & 0 & \tau_3^{(y)} \\ 0 & 1 & 0 & 0 & \tau_1 \\ 0 & \lambda_2 & 0 & 0 & \tau_2 \\ 0 & 0 & 1 & 0 & \tau_3 \\ 0 & 0 & \lambda_4 & 0 & \tau_4 \\ 0 & \tau_3 & \tau_1 & 1 & \tau_1\tau_3 \\ 0 & \tau_4\lambda_2 & \tau_2\lambda_4 & \lambda_2\lambda_4 & \tau_2\tau_4 \end{pmatrix} \begin{pmatrix} \eta \\ \xi_1 \\ \xi_2 \\ \xi_1\xi_2 \\ 1 \end{pmatrix} + \begin{pmatrix} \varepsilon_1 \\ \varepsilon_2 \\ \varepsilon_3 \\ \delta_1 \\ \delta_2 \\ \delta_3 \\ \delta_4 \\ \delta_5 \\ \delta_6 \end{pmatrix},$$

$$\mathbf{y} = \mathbf{\Lambda}_y\boldsymbol{\eta} + \boldsymbol{\varepsilon},$$

$$\begin{pmatrix} \eta \\ \xi_1 \\ \xi_2 \\ \xi_1\xi_2 \\ 1 \end{pmatrix} = \begin{pmatrix} 0 & \gamma_1 & \gamma_2 & \gamma_3 & 0 \\ 0 & 0 & 0 & 0 & 0 \\ 0 & 0 & 0 & 0 & 0 \\ 0 & 0 & 0 & 0 & 0 \\ 0 & 0 & 0 & 0 & 0 \end{pmatrix} \begin{pmatrix} \eta \\ \xi_1 \\ \xi_2 \\ \xi_1\xi_2 \\ 1 \end{pmatrix} + \begin{pmatrix} 0 \\ 0 \\ 0 \\ \phi_{21} \\ 1 \end{pmatrix} 1 + \begin{pmatrix} \zeta \\ \xi_1 \\ \xi_2 \\ \xi_1\xi_2 - \phi_{21} \\ 0 \end{pmatrix}$$

$$\boldsymbol{\eta} = \mathbf{B}\boldsymbol{\eta} + \mathbf{\Gamma}\mathbf{x} + \boldsymbol{\zeta},$$

with

$$\mathbf{\Psi} = \begin{pmatrix} \psi & & & & \\ 0 & \phi_{11} & & & \\ 0 & \phi_{21} & \phi_{22} & & \\ 0 & 0 & 0 & \phi_{11}\phi_{22} + \phi_{21}^2 & \\ 0 & 0 & 0 & 0 & 0 \end{pmatrix},$$

$$\mathbf{\Theta}_\varepsilon = \begin{pmatrix} \theta_1^{(y)} & & & & & & & & \\ 0 & \theta_2^{(y)} & & & & & & & \\ 0 & 0 & \theta_3^{(y)} & & & & & & \\ 0 & 0 & 0 & \theta_1 & & & & & \\ 0 & 0 & 0 & 0 & \theta_2 & & & & \\ 0 & 0 & 0 & 0 & 0 & \theta_3 & & & \\ 0 & 0 & 0 & 0 & 0 & 0 & \theta_4 & & \\ 0 & 0 & 0 & \tau_3\theta_1 & 0 & \tau_1\theta_3 & 0 & \theta_5 & \\ 0 & 0 & 0 & 0 & \tau_4\theta_2 & 0 & \tau_2\theta_4 & 0 & \theta_6 \end{pmatrix}$$

The `LISREL` command file for method WLSA is given in Appendix F.

TABLE 2.5
Results for BEA Model With Two Product Variables

Parameter Estimates			
Parameter	*ML*	*WLS*	*WLSA*
γ_1	–0.11	0.06	0.06
γ_2	1.08	0.93	0.95
γ_3	0.24	0.16	0.17
ψ	0.50	0.58	0.59
R^2	0.54	0.50	0.52
Standard Errors			
Parameter	*ML*	*WLS*	*WLSA*
γ_1	0.16	0.14	0.13
γ_2	0.22	0.16	0.13
γ_3	0.09	0.08	0.09
ψ	0.09	0.09	0.09
T Values			
Parameter	*ML*	*WLS*	*WLSA*
γ_1	–0.64	0.41	0.48
γ_2	5.01	5.93	7.12
γ_3	2.75	2.06	1.83
ψ	5.33	6.45	6.64
Fit Statistics			
Parameter	*ML*	*WLS*	*WLSA*
χ^2	111.78	38.77	45.21
df	29	29	27
P Values	0.00	0.011	0.028
RMSEA	0.10	0.037	0.047

RESULTS

The results are shown in Table 2.5, where only parameters of the structural Equation 1 are included.

Except for chi-square, all three solutions are rather close. Chi-square is much higher for ML than for WLS and WLSA. This is obviously due to the non-normality of the data. The estimate of γ_3 is significant at the 5% level with all the methods although barely so for WLSA. The estimate of γ_1 is not significant with any of the methods whereas the estimate of γ_2 is highly significant with all methods. These results are in line with the results obtained with factor scores. The chi-square goodness-of-fit measures for WLS and

WLSA are significant at the 5% level but not at the 1% level. Using root mean square error of approximation (RMSEA) and the guidelines of Browne and Cudeck (1993), the model would be accepted as being a good approximation to the population covariance matrix.

CONCLUSION

One objective of this chapter was to test empirically the theory of reasoned action as a theory for theoretical construct (latent) variables. It was pointed out that because there are very few observed variables and these are on a low level of measurement, there is very little construct validity for the three construct variables. Nevertheless, the measurement model for the three constructs fitted the data well. Sometime the interaction effect comes out significant, sometimes nonsignificant. This may be due to the small sample. The main conclusion must be that more and better data are needed to test the theory.

Given the data with all their weaknesses, what is the evidence of an interaction effect? Summing up:

1. Grouping on **BE** gives a large difference in slopes that is just significant.
2. Grouping on **VL** gives a large difference in slopes that is highly significant.
3. Using factor scores gives an estimate of γ_3 as 0.09 (0.06) that is not significant.
4. For two product variables both WLS and WLSA give positive estimates of γ_3 that are significant, although just barely so for WLSA.

Altogether these results suggest that there might be an interaction effect. If the sample had been larger this would probably have been detected more clearly.

ACKNOWLEDGMENTS

The research reported in this chapter has been supported by the Swedish Council for Research in the Humanities and Social Sciences (HSFR) under the program *Multivariate Statistical Analysis*.

REFERENCES

Ajzen, I., & Fishbein, M. (1980). *Understanding attitudes and predicting social behavior*. Englewood Cliffs, NJ: Prentice-Hall.

Anderson, T. W. (1984). *An introduction to multivariate statistical analysis* (2nd ed.). New York: Wiley.

Anderson, T. W., & Rubin, H. (1956). Statistical inference in factor analysis. In *Proceedings of the Third Berkeley Symposium* (Vol. V, pp. 111–150). Berkeley: University of California Press.

Bagozzi, R. P. (1986). Attitude formation under the theory of reasoned action and a purposeful behavior reformulation. *British Journal of Mathematical and Statistical Psychology, 25,* 95–107.

Bagozzi, R. P., Baumgartner, H., & Yi, Y. (1992). State versus action orientation and the theory of reasoned action: An application to coupon usage. *Journal of Consumer Research, 18,* 505–518.

Browne, M. W. (1984). Asymptotically distribution-free methods for the analysis of covariance structures. *British Journal of Mathematical and Statistical Psychology, 37,* 62–83.

Browne, M. W., & Cudeck, R. (1993). Alternative ways of assessing model fit. In K. A. Bollen & J. S. Long (Eds.), *Testing structural equation models.* Thousand Oaks, CA: Sage.

Fishbein, M., & Ajzen, I. (1975). *Belief, attitude, intention and behavior: An introduction to research.* Reading, MA: Addison-Wesley.

Graybill, F. A. (1969). *Introduction to matrices with applications in statistics.* Belmont, CA: Wadsworth.

Jöreskog, K. G., & Yang, F. (1996). Nonlinear structural equation models: The Kenny–Judd model with interaction effects. In G. A. Marcoulides & R. E. Schumacker (Eds.), *Advanced structural equation modeling: Issues and techniques* (pp. 57–88). Mahwah, NJ: Lawrence Erlbaum Associates.

Kenny, D. A., & Judd, C. M. (1984). Estimating the nonlinear and interactive effects of latent variables. *Psychological Bulletin, 96,* 201–210.

Koning, R., Neudecker, H., & Wansbeek, T. (1993). Imposed quasi-normality in covariance structure analysis. In K. Haagen, D. Bartholomew, & M. Deistler (Eds.), *Statistical modeling and latent variables* (pp. 191–202). Amsterdam: Elsevier Science Publishers.

Lawley, D. N., & Maxwell, A. E. (1971). *Factor analysis as a statistical method* (2nd ed.). London: Butterworths.

Satorra, A. (1989). Alternative test criteria in covariance structure analysis: A unified approach. *Psychometrika, 54,* 131–151.

Yang Jonsson, F. (1997). *Non-linear structural equation models: Simulation studies of the Kenny–Judd model.* Unpublished doctoral dissertation, Uppsala University, Sweden.

APPENDIX A
PRELIS Input for Computing Mean Vector, Covariance Matrix, and Asymptotic Covariance Matrix

The following `PRELIS` command file reads the raw data on **AA1**, **AA2**, **AA3**, **BE1**, **BE2**, **VL1** and **VL2**, adds the two product variables **BE1VL1** and **BE2VL2**, and computes the sample mean vector $\bar{\mathbf{z}}$, the sample covariance matrix **S**, and the asymptotic covariance matrix **W** needed for ML and WLS. These matrices are saved in files `BEA.ME`, `BEA.CM`, and `BEA.ACC`, respectively.

```
Computing ME, CM, and ACC Matrices
DA NI=7
LA
AA1 AA2 AA3 BE1 BE2 VL1 VL2
```

```
RA=BEA.RAW
CO ALL
NE BE1VL1 = BE1*VL1
NE BE2VL2 = BE2*VL2
OU MA=CM ME=BEA.ME CM=BEA.CM AC=BEA.ACC
```

In a similar way, the following PRELIS input computes the augmented moment matrix **A** and the corresponding asymptotic covariance matrix $\mathbf{W}_a$. These are saved in files BEA.AM, and BEA.ACA, respectively.

```
Computing AM, and ACA Matrices
DA NI=7
LA
AA1 AA2 AA3 BE1 BE2 VL1 VL2
RA=BEA.RAW
CO ALL
NE BE1VL1 = BE1*VL1
NE BE2VL2 = BE2*VL2
OU MA=AM AM=BEA.AM AC=BEA.ACA
```

APPENDIX B
Input Files for Subgroup Analysis

The following PRELIS input file splits the data into two subgroups by using the mean of **BE** = **BE1** + **BE2**. The mean vector, covariance matrix, and asymptotic covariance matrix for those above the mean of BE are saved in file BEABH.ME, BEABH.CM, and BEABH.ACC, respectively. By changing > to <, one can compute the corresponding matrices for those below the mean. In a similar way, one can form subgroups based on **VL** = **VL1** + **VL2**.

```
Selection on BE
DA NI=7
LA
AA1 AA2 AA3 BE1 BE2 VL1 VL2
RA=BEA.RAW
CO ALL
NE BE=BE1+BE2
SD BE1 BE2
SC BE > 10.545
OU MA=CM ME=BEABH.ME CM=BEABH.CM AC=BEABH.ACC
```

To test the equality of γs in Fig. 2.2, the following input file can be used:

```
TESTING MULTIGROUP MODEL ON BE: GROUP LOW
DA NI=5 NO=110 NG=2
LA
AA1 AA2 AA3 VL1 VL2
CM=BEABL.CM
AC=BEABL.ACC
MO NY=3 NX=2 NE=1 NK=1
LE
at
LK
be
FR LY 2 LY 3 LX 2
VA 1 LY 1 LX 1
OU AD=OFF
TESTING MULTIGROUP MODEL ON BE: GROUP HIGH
DA NO=143
CM=BEABH.CM
AC=BEABH.ACC
MO LY=IN LX=IN
OU
```

The preceding input file will produce different γs for both groups. By adding GA=IN to the MO line in the second group, one can get the same γs over groups. Alternatively one can use the following SIMPLIS input:

```
Group Low
TESTING MULTIGROUP MODEL ON BE
Observed variables:
AA1 AA2 AA3 BE1 BE2 VL1 VL2
Covariance Matrix from file BEABL.CM
Asymptotic Covariance Matrix from file BEABL.ACC
Sample size 110
Latent Variables:
vl at
Relationships
AA1 = 1*at
AA2-AA3 = at
VL1 = 1*vl
VL2 = vl
at = vl
Admissibility Check Off

Group High
TESTING MULTIGROUP MODEL ON BE
Covariance Matrix from file BEABH.CM
Asymptotic Covariance Matrix from file BEABH.ACC
```

```
Sample size 143
Relationships

at = vl
Set the error variances of AA1-AA3 free
Set the error variances of VL1-VL2 free
Set the error variance of at free
Set the variance of vl free

End of Problem
```

Specifying the relationship `at = vl` in both groups will give different γs in the groups. By omitting `at = vl` in the second group, the γs will be the same over groups.

APPENDIX C
LISREL Input for Testing Measurement Model

The `LISREL` model in Fig. 2.3 may be tested with the following `LISREL` input:

```
Testing Measurement Model for At Be and Vl
DA NI=7 NO=253
LA
AA1 AA2 AA3 BE1 BE2 VL1 VL2
CM=BEA.CM
AC=BEA.ACC
MO NX=7 NK=3
LK
at be vl
FR LX(2,1) LX(3,1) LX(5,2) LX(7,3)
VA 1 LX(1,1) LX(4,2) LX(6,3)
OU FS ND=3
```

Alternatively, this may be done with the following `SIMPLIS` input file:

```
Testing Measurement Model for At Be and Vl
Observed Variables:
AA1 AA2 AA3 BE1 BE2 VL1 VL2
Covariance Matrix from File BEA.CM
Asymptotic Covariance Matrix from File BEA.ACC
Latent Variables:
At Be Vl
Sample Size: 253
```

```
Equation
   AA1 = 1 * At
   AA2 AA3 = At
   BE1 = 1 * Be
   BE2 = Be
   VL1 = 1 * Vl
   VL2 = Vl
Option: FS ND=3
End of Problem
```

APPENDIX D
PRELIS Input to Compute Factor Scores Covariance Matrix

To compute the covariance matrix of the factor scores, the following PRELIS input may be used:

```
Compute Factor Scores Covariance Matrix
DA NI=7
LA
AA1 AA2 AA3 BE1 BE2 VL1 VL2
RA= BEA.RAW
CO ALL
NE AA1=AA1-4.514
NE AA2=AA2-5.036
NE AA3=AA3-5.004
NE BE1=BE1-5.115
NE BE2=BE2-5.340
NE VL1=VL1-5.814
NE VL2=VL2-5.723
NE AT=.195*AA1+.303*AA2+.333*AA3+.009*BE1+.023*BE2+.053*VL1+0.18*VL2
NE BE=.010*AA1+.015*AA2+.017*AA3+.188*BE1+.474*BE2+.110*VL1+.037*VL2
NE VL=.010*AA1+.015*AA2+.017*AA3+.019*BE1+.048*BE2+.656*VL1+.222*VL2
NE BEVL=BE*VL
SD AA1-VL2
OU MA=CM CM=BEAFS.CM AC=BEAFS.ACC
```

APPENDIX E
LISREL Input for Model With Two Product Variables, ML and WLS

To estimate the model using ML estimation, the following LISREL input file can be used. This file uses nonlinear constraints, such as the variance of the latent product variable equals the variance of the first latent variable times the variance of the second latent variable plus the squared covariance of the latent variables. These constraints are specified in the lines starting with CO.

```
BEA Model with two product variables
Fitting Model to Mean Vector and Covariance Matrix by ML
DA NI=9 NO=253
LA
AA1 AA2 AA3 BE1 BE2 VL1 VL2 BE1VL1 BE2VL2
CM=BEA.CM
ME=BEA.ME
MO NY=3 NX=6 NE=1 NK=3 TD=SY TY=FR TX=FR KA=FR
FR LY(2) LY(3) GA(1) GA(2) GA(3) LX(2,1) LX(4,2) PH(1,1)-PH(2,2)
FI PH(3,1) PH(3,2)
VA 1 LY(1) LX(1,1) LX(3,2) LX(5,3)
FI KA(1) KA(2)
CO LX(5,1)=TX(3)
CO LX(5,2)=TX(1)
CO LX(6,1)=TX(4)*LX(2,1)
CO LX(6,2)=TX(2)*LX(4,2)
CO LX(6,3)=LX(2,1)*LX(4,2)
CO PH(3,3)=PH(1,1)*PH(2,2)+PH(2,1)**2
CO TD(5,1)=TX(3)*TD(1,1)
CO TD(5,3)=TX(1)*TD(3,3)
CO TD(5,5)=TX(1)**2*TD(3,3)+TX(3)**2*TD(1,1)+PH(1,1)*TD(3,3)+ C
   PH(2,2)*TD(1,1)+TD(1,1)*TD(3,3)
CO TD(6,2)=TX(4)*TD(2,2)
CO TD(6,4)=TX(2)*TD(4,4)
CO TD(6,6)=TX(2)**2*TD(4,4)+TX(4)**2*TD(2,2)+LX(2,1)**2*PH(1,1)
   *TD(4,4)+ C LX(4,2)**2*PH(2,2)*TD(2,2)+TD(2,2)*TD(4,4)
CO KA(3)=PH(2,1)
CO TX(5)=TX(1)*TX(3)
CO TX(6)=TX(2)*TX(4)
ST 5 TE(1)-TE(3) TD(1)-TD(4)
ST 5 TY 1 - TY 3 TX 1 - TX 4
ST 1 GA 1 - GA 3
OU AD=OFF EP=0.001 IT=500
```

The same file can be used for the WLS estimation method, but the line

```
AC=BEA.ACC
```

must be added after the line

```
ME=BEA.ME
```

APPENDIX F
LISREL Input for Model With Two Product Variables, WLSA

Because $\mathbf{W}_a$ is singular, LISREL automatically computes and uses the generalized inverse $\mathbf{W}^{-}$ and adjusts the degrees of freedom.[2] Because Ψ is singular, one must put AD=OFF on the OU line. There is no reference variable for the fourth factor, so starting values must be provided and NS must be specified on the OU line. It is also a good idea to put SO on the OU line to tell LISREL not to check the scales for the factors.

```
BEA Model with two product variable
Fitting Model by WLS using Augmented Moment Matrix
DA NI=10 NO=253
LA
AA1 AA2 AA3 BE1 BE2 VL1 VL2 BE1VL1 BE2VL2 CONST
CM=BEA.AM
AC=BEA.ACA
MO NY=9 NE=5 NX=1 BE=FU GA=FI TE=SY PS=SY,FI FI
FR LY(2,1) LY(3,1) LY(5,2) LY(7,3) BE(1,2) BE(1,3) BE(1,4)
FR LY(1,5) LY(2,5) LY(3,5) LY(4,5) LY(5,5) LY(6,5) LY(7,5)
FR PS(1,1) PS(2,2) PS(3,2) PS(3,3)
VA 1 LY(1,1) LY(4,2) LY(6,3) LY(8,4) GA(5)
CO LY(8,2)=LY(6,5)
CO LY(8,3)=LY(4,5)
CO LY(8,5)=LY(4,5)*LY(6,5)
CO LY(9,2)=LY(7,5)*LY(5,2)
CO LY(9,3)=LY(5,5)*LY(7,3)
CO LY(9,4)=LY(5,2)*LY(7,3)
CO LY(9,5)=LY(5,5)*LY(7,5)
CO GA(4)=PS(3,2)
CO PS(4,4)=PS(2,2)*PS(3,3)+PS(3,2)**2
CO TE(8,4)=LY(6,5)*TE(4,4)
CO TE(8,6)=LY(4,5)*TE(6,6)
CO TE(8,8)=LY(4,5)**2*TE(6,6)+LY(6,5)**2*TE(4,4)+PS(2,2)*TE(6,6)+ C
PS(3,3)*TE(4,4)+TE(4,4)*TE(6,6)
CO TE(9,5)=LY(7,5)*TE(5,5)
CO TE(9,7)=LY(5,5)*TE(7,7)
CO TE(9,9)=LY(5,5)**2*TE(7,7)+LY(7,5)**2*TE(5,5)+LY(5,2)**2*PS(2,2)
           *TE(7,7)+ C LY(7,3)**2*PS(3,3)*TE(5,5)+TE(5,5)*TE(7,7)
ST 1.03 LY(2,1)
ST 1.14 LY(3,1)
```

[2]LISREL 9 does this automatically; with LISREL 8 one must first run GINV to produce the generalized inverse, save it as BEA.WMC, and then read this into LISREL with the line WM=BEA.WMC.

```
ST 1.26 LY(5,2)
ST 0.89 LY(7,3)
ST -.04 BE(1,2)
ST 1.01 BE(1,3)
ST 0.12 BE(1,4)
ST 5 LY(1,5) LY(2,5) LY(3,5) LY(4,5) LY(5,5) LY(6,5) LY(7,5)
ST 0.38 PS(1,1)
ST 0.64 PS(2,2)
ST 0.44 PS(3,2)
ST 0.60 PS(3,3)
ST 0.5 TE(1,1) TE(2,2) TE(3,3) TE(4,4) TE(5,5) TE(6,6) TE(7,7)
OU NS AD=OFF SO
```

3

Modeling Interaction and Nonlinear Effects With Mx: A General Approach

Michael C. Neale
Medical College of Virginia

This chapter has three primary goals. First, it introduces a new, simple, and general way to express the constraints in nonlinear and interactive models of the Kenny and Judd (1984) type. Only two matrix constraints are needed for all two-way interaction product variable models. Second, two Kenny and Judd models are implemented with the structural equation modeling program Mx. Third, the alternative, multigroup approach to testing for moderator variables is expanded to cover the general case of continuous moderator variables, requiring a large number of groups. The Mx scripts and diagrams presented and the program itself are available on the Internet for free download.

The linear model, being powerful, general, and simpler than nonlinear models, has been widely used in statistical research. In the behavioral sciences, theories and hypotheses are often relatively simple, so that linear relationships between variables are of primary interest. When this initial simplicity is found to be inadequate to explain the data, or when theories develop, hypotheses about more complex relationships arise. Although it is quite straightforward to add nonlinear or "interaction" terms to multiple regression using product variables, it is much more difficult to do so with latent variables. The goals of this chapter are to present general specifications for nonlinear latent variable models and to implement them using Mx software.

The limitations of linear modeling have long been recognized. As Lawley and Maxwell (1963) pointed out, Bartlett (1953) suggested amending the basic factor model to include second-order and product terms. For example, a two-factor model of the form:

$$x = l_1 f_1 + l_2 f_2 + e$$

might become

$$x = l_1 f_1 + l_2 f_2 + l_3 f1^2 + l_4 f2^2 + l_5 f1 f2 + e$$

where the l_3 to l_5 are the loadings on the two-way nonlinear and interactive components of the model. Etezadi and colleagues described some early implementations of nonlinear factor models (Etezadi-Amoli & Ciampi, 1983; Etezadi-Amoli & McDonald, 1983). As is clear from their work and that of Kenny and Judd (1984), the statistical modeling of these nonlinear effects is not straightforward within latent variable modeling. In particular, parameter constraints in these models are complex, awkward to specify, and can change dramatically when relatively minor modifications are made to the linear part of the model.

In this chapter, I describe a general matrix form for the constraints associated with nonlinear factor models, which greatly simplifies their specification. Being general, they do not require any changes for particular applications. The Kenny and Judd (1984) models are implemented with the computer program Mx (Neale, 1997) for illustration.

LATENT VARIABLE INTERACTION MODELS

Kenny and Judd (1984) described simple interaction models, which are shown as path diagrams in Figs. 3.1 and 3.2. In Fig. 3.1, the latent factor *XX* represents the quadratic effect of the latent factor *X*, being the multiplicative product of *X* with *X*. In Fig. 3.2, the latent factor *XZ* represents the multiplicative interaction between the latent factors *X* and *Z*. The effects of the nonlinear latent factors *XX* and *XZ* are identified through two external sources of information: (a) the two-way cross-products of the observed variables (X_1X_2, X_1Z_2, etc.), and (b) an external indicator *Y* related to the nonlinear factors.

The variances and covariances of the nonlinear factors are functions of the variances and covariances of the linear latent variables; they form one set of constraints. The path coefficients from the nonlinear factors to the cross-products of the observed variables are functions of the path coefficients from the linear factors to the observed variables, and form the second set of constraints. Both sets may be derived by taking expectations as described by Kenny and Judd (1984) and Kendall and Stuart (1977). These results are not reproduced here; instead I give a conceptual overview of the origin and form of the constraints.

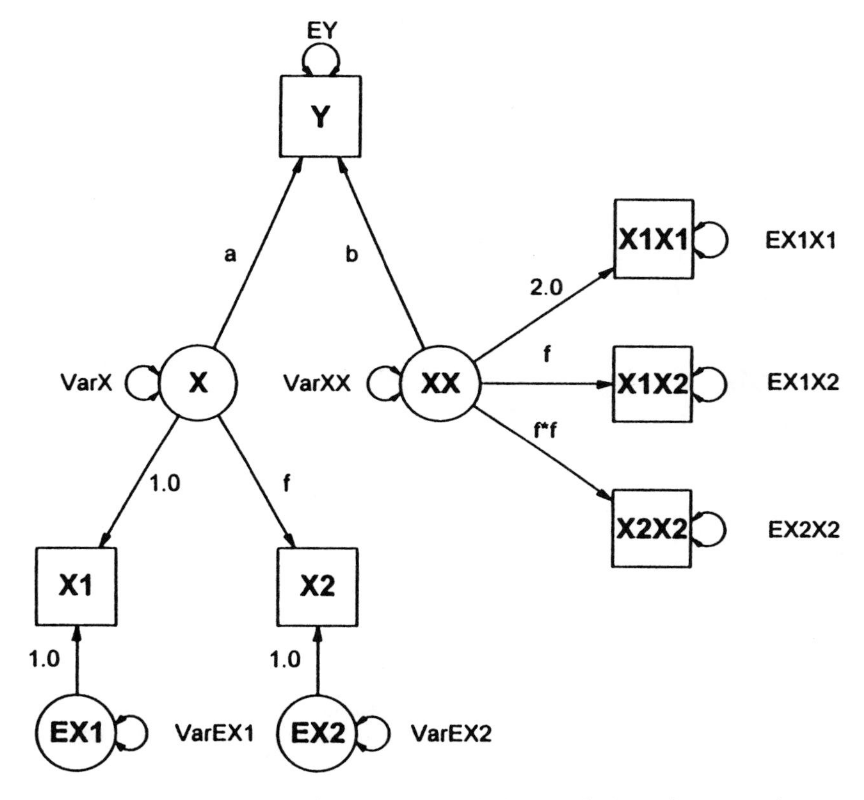

FIG. 3.1. Kenny and Judd nonlinear model, showing the latent factor X and two observed indicators $X1$ and $X2$. The quadratic latent factor XX impacts the product observed variables $X1X1$ $X1X2$ and $X2X2$ and the indicator Y.

Constraints on the Covariances of Product Variables

In Fig. 3.2, the observed variable $X1$ is the linear sum of the effects of the factor X and its error E_{X1}. Similarly, $Z1$ is the sum of the effects of its factor, Z, and its error, E_{Z1}. The variables $X1$ and $Z1$ are called *constituent variables,* whose *product variable* $X1Z1$ may be written:

$$X1Z1 = (aX + E_{X1})(bZ + E_{Z1}).$$

The variance of this product variable is:

$$\begin{aligned}\mathrm{Var}(X1Z1) = {} & a^2b^2\mathrm{Var}(XZ) + a^2\mathrm{Var}(X)\mathrm{Var}(E_{Z1}) + \\ & b^2\mathrm{Var}(Z)\mathrm{Var}(E_{X1}) + \mathrm{Var}(E_{X1})\mathrm{Var}(E_{Z1}). \end{aligned} \tag{1}$$

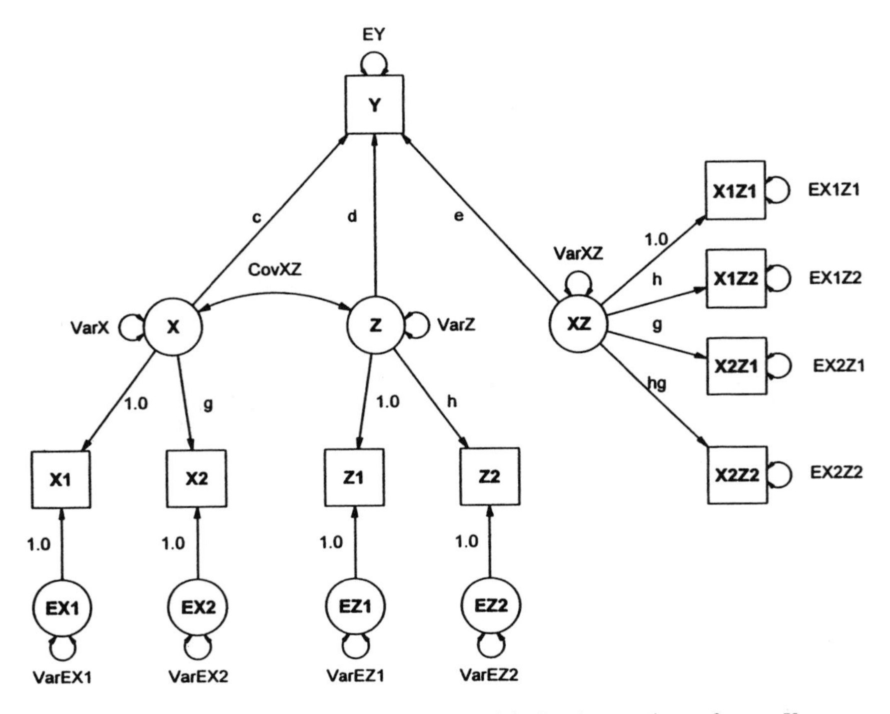

FIG. 3.2. Kenny and Judd interaction model, showing two latent factors X and Z each with two observed indicators (subscripted 1 and 2). The product latent factor XZ impacts the product observed variables X_iZ_i and the indicator Y.

The most important insight here is that the variance of the product variable $X1Z1$ is a known function of the interaction latent variable XZ. If another variable, say Y, is an indicator of XZ then it should covary with the product variable $X1Z1$ (and all other variables that are the product of an indicator of X and an indicator of Y). The extent of this covariance uniquely identifies the factor loading of Y on XZ because the loadings of $X1$ and $Z1$ on XZ are simple constraints of parameters that are identified from the noninteractive part of the model.

Under multivariate normality, if the constituent variables have zero mean (i.e., they are "centered"), all product variables $X_iX_j|i \neq j$ formed from them will have zero means (Kendall & Stuart, 1977). Furthermore, the covariance between product variables and the constituent variables is zero. The variances and covariances of the product variables are functions of the variances and covariances of their constituent variables. For example,

$$\text{Var}(X^2) = 2(\text{Var}(X))^2, \tag{2}$$

$$\text{Var}(XY) = (\text{Var}(X))(\text{Var}(Y)) + \text{Cov}(XY)^2 \tag{3}$$

(Kenny & Judd, 1984). These constraints may be derived individually from expectations of the fourth moments, which is a straightforward but tedious and error-prone task. As Jöreskog and Yang (1996) noted, "Utmost care must be taken to specify the constraints in the model correctly. A single mistake may have severe consequences . . . these constraints are quite complex and it is easy to make a mistake" (p. 85). The solution to this problem is to express all the constraints in general matrix form, thus reducing the constraints to two matrix expressions. These constraints are invariant to changes in the number of variables in the model, or in the structural relations between them.

Extrapolating from the results of Henderson and Searle (1979) and Searle (1982), we can write the covariance matrix of the cross products of a vector $\mathbf{v} = v_1 \ldots v_k$ of variables with covariance matrix $\mathbf{S}$ as

$$\text{Cov}(\mathbf{v} \otimes \mathbf{v}) = (\mathbf{S} \otimes \mathbf{S})(\mathbf{I} + \mathbf{J}) \tag{4}$$

where $\mathbf{I}$ is the identity matrix of order k^2 and $\mathbf{J}$ is a permuted identity matrix such that

$$\mathbf{J}\text{Vec}(\mathbf{A}) = \text{Vec}(\mathbf{A}')$$

where the Vec operator stacks the columns of a matrix one under the other to form a single-column vector (Searle, 1982). For example, with

$$\mathbf{A} = \begin{bmatrix} 1 & 3 & 5 \\ 2 & 4 & 6 \end{bmatrix}$$

we have

$$\text{Vec}(\mathbf{A}) = \begin{bmatrix} 1 \\ 2 \\ 3 \\ 4 \\ 5 \\ 6 \end{bmatrix} \text{Vec}(\mathbf{A}') = \begin{bmatrix} 1 \\ 3 \\ 5 \\ 2 \\ 4 \\ 6 \end{bmatrix}$$

so

$$\mathbf{J} = \begin{bmatrix} 1 & 0 & 0 & 0 & 0 & 0 \\ 0 & 0 & 1 & 0 & 0 & 0 \\ 0 & 0 & 0 & 0 & 1 & 0 \\ 0 & 1 & 0 & 0 & 0 & 0 \\ 0 & 0 & 0 & 1 & 0 & 0 \\ 0 & 0 & 0 & 0 & 0 & 1 \end{bmatrix}.$$

Constraints on Product Variable Factor Loadings

The expansion of the expression for the variance of a product term, as shown for the $X1Z1$ case in Equation 1, follows a regular form. This regularity is slightly obscured in Equation 1 because the loading of $X1$ on its error is fixed at 1.0, as is the loading of $Z1$ on its error. The regularity comes from multiplying every term of the expanded expression for $X1$ by every term of the expanded expression for $Z1$, which is succinctly written as a Kronecker product for the general case of a vector of variables $\mathbf{w} = \mathbf{bv}$:

$$\mathbf{w} \otimes \mathbf{w} = (\mathbf{b} \otimes \mathbf{b})(\mathbf{v} \otimes \mathbf{v}).$$

Using the expression for $\text{Cov}(\mathbf{v} \otimes \mathbf{v})$ from Equation 4 we can write

$$\text{Cov}(\mathbf{w} \otimes \mathbf{w}) = (\mathbf{b} \otimes \mathbf{b})\,[(\mathbf{S} \otimes \mathbf{S})(\mathbf{I} + \mathbf{J})]\,(\mathbf{b} \otimes \mathbf{b})',$$

which is a relatively straightforward expansion of the expression for the constituent variables:

$$\text{Cov}(\mathbf{w}) = \mathbf{bSb}'.$$

Therefore, we can write a partitioned matrix form for the covariance of the constituent variables $\mathbf{w}$, the product variables $\mathbf{w} \otimes \mathbf{w}$, and external indicator variables $\mathbf{y}$:

$$\text{Cov}(\mathbf{w} : \mathbf{w} \otimes \mathbf{w} : \mathbf{y}) = \mathbf{R} = \left[\begin{array}{c|c|c} \mathbf{bSb}' & \mathbf{0} & \mathbf{bSp}' \\ \hline \mathbf{0} & \mathbf{mCm}' & \mathbf{mCq}' \\ \hline \mathbf{pSb}' & \mathbf{qCm}' & \mathbf{pSp}' + \mathbf{qCq}' + \mathbf{E} \end{array}\right] \qquad (5)$$

where:

$\mathbf{m} = \mathbf{b} \otimes \mathbf{b}$,
$\mathbf{C} = (\mathbf{S} \otimes \mathbf{S})(\mathbf{I} + \mathbf{J})$,
$\mathbf{p}$ = Loadings of $\mathbf{y}$ on constituent latent variables,
$\mathbf{q}$ = Loadings of $\mathbf{y}$ on product latent variables,
$\mathbf{E}$ = Residual variance of $\mathbf{y}$ variables.

Using Kronecker products generates redundancy in the two-way product variables because, for example, both $X1X2$ and $X2X1$ are created. The duplicated variables would correlate perfectly in a data set—they add no new information—and this would cause problems for fitting functions such as maximum likelihood (ML). To eliminate the redundant product variables from the predicted covariances, a filter matrix **F** can be constructed to select only the constituent variables, the indicator variables, and a subset of the product variables for analysis. **F** is an elementary matrix with columns corresponding to all the rows of **R** and rows indicating the subset of variables to be analyzed; it has a one in elements where the row variable is the same as the column variable and a zero otherwise. The predicted covariance structure for the subset of variables is then $\mathbf{\Sigma} = \mathbf{FRF}'$.

IMPLEMENTATION WITH Mx

Mx (Hamagami, 1997; Neale, 1997) is a software package for structural equation and other mathematical modeling. It combines a matrix algebra interpreter with numerical optimization software to facilitate the implementation of complex mathematical models, and offers several ways to implement nonlinear constraints among the parameters. Models are specified with the Mx script language which takes the general form of declaring matrices, setting matrix elements to certain values, setting matrix elements to be fixed or free parameters, and using these matrices in matrix algebra statements to define the predicted covariance and mean structure under the model.

Kenny and Judd Nonlinear Model

Appendix A shows the Mx script for the Kenny and Judd (1984) nonlinear model; it is internally commented (everything after a ! is a comment) and is somewhat self-explanatory, but several features are worthy of note. The first few lines define a number of variables, such as `nfac`, that are used to set the dimensions of matrices used to specify the model. In contrast to the usual use of the number of factors, `nfac` refers to the number of factors *including the error terms for the variables* $X1$ *and* $X2$. The number of observed variables `nvar` does not include the number of indicator variables, `nindic`. Following the `#define` statements the script reads the title, variable labels, and data. The `Begin Matrices` section uses the `#define` variables to declare matrix dimensions and does not need to be modified for other models.

Specification of the linear part of the model for the $X1$ and $X2$ variables is accomplished by setting parameters into the matrices **b** and **S**. For the Y variable the linear loadings are specified in the matrix **p** and the nonlinear loadings are specified in **q**. The filter matrix **F** must be modified when the number of observed variables changes, or when different product variable indicators are used. The matrix **J** is the reordered identity matrix described previously; for models with different numbers of factors, a different size **J** matrix must be used. These matrices can be quite large and tedious to generate, so a simple FORTRAN program makej.f was written and is freely available on the Mx website at `http://griffin.vcu.edu` in the examples directory under Kenny and Judd, along with Appendices A and B and the data files.

The remaining parts of the model specification, including all the effects on the product variables and their covariances with the Y variable(s), are automatically generated in the matrix algebra section, and are therefore invariant across scripts. The nine parts of the partitioned matrix **R** of Equation 5 are concatenated in the `Covariance` statement. Option GLS requests generalized least squares (GLS) as the fitting function.

Results from fitting the model give a fit of 33.15 with 14 *df* (degrees of freedom), which indicates significant departure of the observed data from those predicted by the model (p = .003). RMSEA is .05, which indicates quite a good fit. Using the maximum likelihood fit function, a chi-squared of 19.13 is obtained, which is an acceptable fit (p = .016), and the RMSEA is .03—a good fit. It is interesting that GLS and ML fit functions are quite divergent for these data, despite a moderately large sample size. Inherent non-normality of the data may be responsible; the observed product variables cannot be normally distributed, nor can the indicator variable Y to the extent that it is a function of the product factor XX. Jöreskog and Yang (1996) discussed the use of asymptotic weighted least squares to handle non-normality in these models, but this too is problematic, because the weight matrices are less than full rank. Possibly, a ML fit function could be devised that either operates with the raw data, or uses higher moments than means and covariances (e.g., Bentler, 1983) to describe the data.

Parameter estimates from the nonlinear Mx model are close to the values used by Kenny and Judd (1984) to simulate the data, and are close to the estimates reported in the original article. They are not exactly the same, perhaps because of minor optimization or specification differences. The GLS estimates from the script in Appendix A are: a = .248; b = –.502; f = 0.626; $EX1$ = .163; $EX2$ = .554; EY = .196; Var$X$ = .980.

Kenny and Judd Interaction Model

Appendix B shows the script for the Kenny and Judd (1984) interaction model, which closely follows the format for the nonlinear model described previously. The changes from the nonlinear model specification are as follows:

1. The number of factors is six and observed variables is four, leading to several changes in the `#define` statements.
2. The labels of the observed variables are different.
3. The CMatrix data file has changed.

4. The model specification in **b** and **S** is different.
5. A larger matrix **J** is read.
6. A different filter matrix **F** selects a subset of product variables from the full set of all two-way products.
7. The model specification for the *Y* variable in matrices **p** and **q** differ.
8. Matrices **b**, **S**, and **F** have different labels.

The GLS function for the interaction model is 100.06 with 32 *df*, with RMSEA of .065—a poor fit (p = .000), if the GLS is to be interpreted as χ^2. Again the ML fit function suggests a much better fit than GLS, being 44.79 (p = .066), giving RMSEA of 0.028—a good fit. GLS parameter estimates are: g = .643; h = .667; c = –0.167; d = 0.325; e = 0.705; Var$X$ = 1.816; Var$Z$ = 1.524; CovXZ = 0.352; E_{X1} = .421; E_{X2} = .721; E_{Z1} = .429; E_{Z2} =.576; E_Y = .299. These estimates are very close to those reported by Kenny and Judd (1984).

DISCUSSION

General Constraints

Although nonlinear factor models were implemented almost 20 years ago, they have been rarely used since then. A small resurgence of development and application has recently occurred, with articles by Ping and Jöreskog and others being especially notable. The need for detailed understanding of the method, and the painstaking specification of the nonlinear constraints required to implement models even slightly different from those published, have conspired to make nonlinear factor models relatively unpopular. Considering the enormous growth of (linear) structural equation modeling during the same period, nonlinear modeling is sparse indeed. It is hoped that the general method presented here will lift the barrier to research using nonlinear models. Essentially, the problem has been reduced to the specification of the linear part of the model.

The Kenny and Judd (1984) models contain only first-order factors, so that the elements of **b** and **p** are parameters or zero elements. For more general treatment, we could redefine the model specification using RAM notation (McArdle & McDonald, 1984) within the constituent variable part so that **S** contains all two-headed paths, and **B** contains all single-headed paths. The matrix **b** would then be replaced by $\mathbf{I} - \mathbf{B}^{-1}$. Similar substitutions could be made for the **y** part of the model.

In principle, extension to higher-order products is simple. However, such statistics depend on sixth, eighth, and even higher moments, which in practice would require extremely large sample sizes to estimate parameters of a model that depended on them.

A serious statistical problem with these methods still remains: the non-normality of the data. With multivariate normal constituent variables, all product variables will be markedly non-normal. Any variable that is a function of a nonlinear factor (be it an interaction between variables XZ type, quadratic X^2 type, or higher-order interaction) will also exhibit non-normality. Thus with either ML or GLS, the goodness-of-fit statistics will be biased and likelihood-ratio fit statistics based on the differences between these statistics will be inaccurate. This difficulty is only partially solved by the use of asymptotic least squares, as pointed out by Jöreskog and Yang (1996), because the weight matrix computed from constituent and product variables is not of full rank.

Multiple Group Approaches to Nonlinear Modeling

As discussed by Rigdon et al. in chapter 1, the product variable approach to modeling interactions may not be the best or the most versatile method. Much more straightforward and flexible is the multigroup structural equation modeling method. Different levels of the interaction variable are used to classify subjects into different groups. This procedure is simple to implement if the interaction variable has only a few levels, but may run foul of non-positive definite covariance matrices when the number of subjects in a group is small. Mx has powerful features for multigroup modeling because it was developed with a special view to facilitating genetic modeling using different types of relatives, such as identical and fraternal twins. In addition, the ability to fit models to raw data allows a simple extension to the case where the number of groups is very large and the number of subjects within each group is as few as one.

To illustrate the application of the multiple group method with Mx, the example of MacKenzie and Spreng (1992) is used. This example is discussed in more detail in chapter 1 of this volume, so only a brief description is given here. The primary research questions of the study concerned the moderating effects of motivation on the relationship between factors representing attitude to brands and to their advertisements. Path diagrams of the models for the two groups are shown in Fig. 1.2. The models are designed to test whether the high- and low-motivation samples (groups 1 and 2) differ in their factor means, or in the size of the path from the advertisement attitude latent variable to the brand attitude latent variable, or both. Tests of these hypotheses are achieved by fitting four models: (a) neither components differ between groups; (b) factor means differ; (c) the Ad $\rightarrow$ Brand path differs; and (d) both components differ. Significant improvement in fit of a model in which one or both components differ would indicate a moderating effect of high versus low motivation. An Mx script for the MacKenzie and Spreng (1992) model is shown in Appendix C and is available on the Internet at `http://views.vcu.edu/mx/examples.html` in the MacKenzie link as "mackenzie.mx." The script contains a multiple-fit option, which sequentially fits all four models. The results obtained with Mx closely match those obtained using LISREL and AMOS in chapter 1. For those who prefer not to use the Mx script language, the same results can be obtained using the Mx Windows graphical user interface by drawing separate diagrams for groups 1 and 2, and fitting the model directly from the diagrams, as one would do with AMOS. The relevant diagram-based model and all data files may be downloaded from the same area of the Mx website in mackenzie.zip, and the graphical interface itself may be obtained from `http://www.vipbg.vcu.edu/mxgui/`.

Continuous interaction variables or *moderation* can be specified through definition variables in Mx (Neale, 1997) using the raw data. Essentially, the observed moderator variable will be used as a parameter in the model. Therefore, the predicted covariance matrix and means may be different for every subject in the sample, which is the natural extension of allowing the predicted covariances and means to differ between groups. This general approach subsumes the special case of a limited number of groups through the use of a dummy group variable that codes for group membership in the sample. A full description of these methods is beyond the scope of this chapter, but an example script and an example diagram file may be found at the website in the example files `contin.mx` and `contin.mxd`. Figure 3.3 shows the example diagram, which contains the usual mixture of latent variables (circles) and observed variables (squares). Less familiar are the triangles, which contribute only to the means of the variables, and not to their variances or covariances. Least familiar of all are the diamonds, which contain the observed value of group. Suppose that group always takes the value 1 for those subjects in the high group, and 0 for subjects in the low group. The path Z-low therefore is the only path from latent factor Ad to Brand, as the product of Group (0) by A3 is 0. In the high group, however, there are two paths from Ad to Brand, as the value of Group is 1. Hence, the path A3 contains the *deviation* of the high group's Ad-to-Brand path from the value of this path in the low group. Similar effects are taking place for the factor means; the mean of Ad in the low group is simply $Y2$, whereas in the high group it is $X2 + Y2$. If the variable Group was a continuous measure, no changes to the model would be required. The factor mean of Ad would simply be $Y2$ + Group*$X2$. More complicated nonlinear effects of a continuous Group variable might be modeled by putting Group on two consecutive paths, so that the square of the Group variable moderates the relationship.

Although they are promising, the continuous moderator variable methods have their limitations. First, having a different model for each subject in the

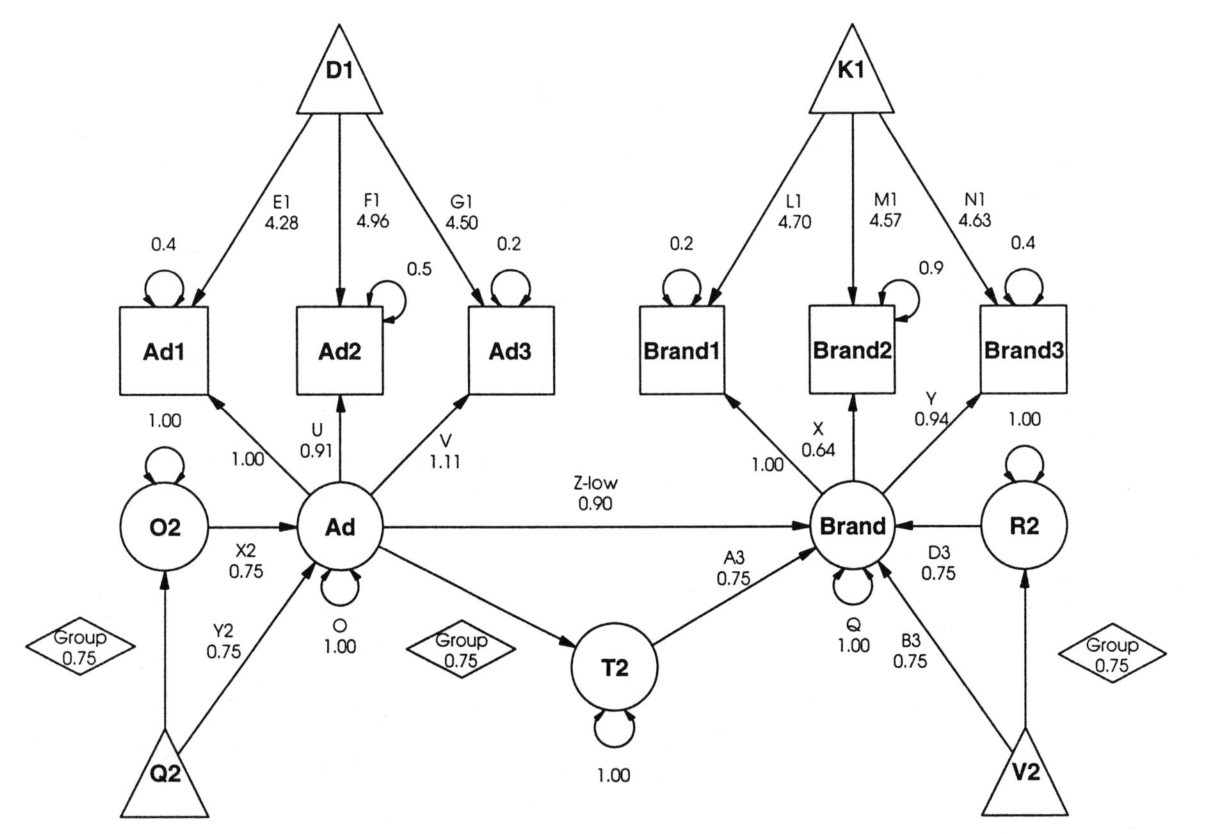

FIG. 3.3. MacKenzie and Spreng interaction model, showing two latent factors *Ad* and *Brand* each with three observed indicators (subscripted 1, 2, and 3). Ad causes Brand through two pathways, one direct (Z-low) and one indirect (Group times A3). Similar direct and indirect paths are specified from the mean variables (triangles) to the latent factors, to model factor means that vary as a function of Group. Group may be dummy coded as 1 or 0 to emulate a two-group model, or may be a continuous measure of motivation to model a simple linear interaction. In both cases, the model must be fitted to the raw data.

sample can dramatically increase computer time required to fit the model, especially when sample size N is large, because there are N different predicted covariance matrices to invert to compute the likelihood. Second, there is an increased chance that an infeasible model, such as one where the predicted covariance matrix is ill-conditioned or not positive definite, will be generated. Such problems make working with continuous moderator variables more sensitive to starting values than usual. Third, the methods are limited to data that are (approximately) normally distributed conditional on the value of the interaction variable. This limitation may in fact be less restrictive than the usual assumption that the data are multivariate normal, for the data need only be multivariate normal conditional on the values of the moderator variables. An example where this limitation is effectively a relaxation of the usual multivariate normality assumption is in the two-group case when the groups are drawn from two multivariate normal distributions with different means. The joint distribution of both groups would not be multivariate normal, although the distribution of each group considered separately would be. Finally, it is important to establish that the moderator variables do not have a main effect on the variables being analyzed, as this can bias the detection of moderating effects. The most practical way to remove the main effects of the moderators may be to analyze the residuals of the variables with the moderators' main effects regressed out.

In conclusion, both the indicant product method of Kenny and Judd and the multigroup structural equation modeling approach to moderating variables have been simplified and generalized. It is hoped that the methods presented in this chapter will open doors to new areas of research and application of models for moderating and nonlinear interactions.

ACKNOWLEDGMENTS

The author was supported by NIH grants MH-40828, HL48148, MH45268, MH41953, RR-08123, and AA-09095.

REFERENCES

Bartlett, M. S. (1953). *Factor analysis in psychology as a statistician sees it* (Vol. 3, *Nordisk Psykologi's* Monograph Series). Copenhagen: Ejnar Mundsgaards.

Bentler, P. M. (1983). Some contributions to efficient statistics for structural models: Specification and estimation of moment structures. *Psychometrika, 48,* 493–517.

Etezadi-Amoli, J., & Ciampi, A. (1983). Simultaneous parameter estimation for the multiplicative multiattribute utility model. *Organizational Behavior and Human Performance, 32,* 232–248.

Etezadi-Amoli, J., & McDonald, R. P. (1983). A second generation nonlinear factor analysis. *Psychometrika, 48*(3), 315–342.

Hamagami, F. (1997). A review of the Mx computer program for structural equation modeling. *Structural Equation Modeling, 4*(2), 157–175.

Henderson, H. V., & Searle, S. R. (1979). Vec and vech operators for matrices, with some uses in jacobians and multivariate statistics. *Canadian Journal of Statistics, 7,* 65–81.

Jöreskog, K., & Yang, F. (1996). Nonlinear structural equation models: The Kenny–Judd model with interaction effects. In G. Marcoulides & R. E. Schumacker (Eds.), *Advanced structural equation modeling: Issues and techniques* (pp. 57–88). Mahwah, NJ: Lawrence Erlbaum Associates.

Kendall, M., & Stuart, A. (1977). *The advanced theory of statistics.* New York: Macmillan.

Kenny, D. A., & Judd, C. M. (1984). Estimating the nonlinear and interactive effects of latent variables. *Psychological Bulletin, 96,* 201–210.

Lawley, D. N., & Maxwell, A. E. (1964). *Factor analysis as a statistical method.* London: Butterworths.

McArdle, J. J., & McDonald, R. P. (1984). Some algebraic properties of the reticular action model. *British Journal of Mathematical and Statistical Psychology, 37,* 234–251.

Neale, M. C. (1997). *Mx: statistical modeling* (4th ed.). Richmond, VA: Author.

Searle, S. R. (1982). *Matrix algebra useful for statistics.* New York: Wiley.

APPENDIX A
Mx Script for Nonlinear Effects

```
Title Kenny & Judd Psychol Bull 1984 nonlinear model. General form
 Data NGroups=1 NI=ninput NO=500
 CMatrix File=kennynon.cov
 Labels X1 X2 X1^2 X2^2 X1*X2 Y
 Select X1 X2 X1^2 X1*X2 X2^2 Y ;
 Begin Matrices;
  B Full nvar nfac           ! Loadings on Linear factors (including
                             !  error terms)
  S Diag nfac nfac           ! Covariances of Linear factors
  I Iden nfacsq nfacsq       ! Identity matrix
  J Full nfacsq nfacsq       ! Reordering matrix
  P Full nindic nfac         ! Paths from Linear factors to indica-
                             tor variables
  Q Full nindic nfacsq       ! Paths from Squared factors to
                             ! indicator variables
  E Diag nindic nindic Free  ! Residual variance of indicator variables
  F Full ninput ntot         ! Filter matrix to select some of the
                             ! variables
  Z Zero nvar nvarsq         ! Zeroes
End Matrices;

Specify B
 10 0 0
```

```
  0 13 14
 Matrix B
  .5 1 0
  0 .5 .5
 Labels Row B X1 X2
 Labels Col B ErrorX1 Factor1 ErrorX2

 Free S 2 2
 Matrix S
  1 .2 1
 Labels Row S ErrorX1 Factor1 ErrorX2
 Labels Col S ErrorX1 Factor1 ErrorX2

 Matrix E
  1
 Label Col E YErrorVar

 Matrix F
  1 0 0 0 0 0 0
  0 1 0 0 0 0 0
  0 0 1 0 0 0 0
  0 0 0 1 0 0 0
  0 0 0 0 0 1 0
  0 0 0 0 0 0 1

 Label Col F X1 X2 X1^2 X1*X2 X2*X1 X2^2 Y
 Label Row F X1 X2 X1^2 X1*X2 X2^2 Y
 Matrix J File=j3.mat

 Free Q 1 5
 Label Row Q Y
 Label Col Q E1^2 E1F1 E1E2 F1E1 F1^2 F1E2 E2E1 E2F1 E2^2

 Free P 1 2
 Label Row P Y
 Label Col P ErrorX1 Factor1 ErrorX2

 Begin Algebra;
  C = D@D*(I+J) ; ! Covariances among combinations
  M = B@B ;       ! Loadings on combinations
 End Algebra;
 Covariance F*( b*S*b' |      Z | b*S*p' _
                    Z' | m*C*m'| m*C*q' _
               p*S*b' | q*C*m'| p*S*p' + q*C*q' + E ) *F';
 Option GLS
End Group;
```

APPENDIX B
Mx Script for Interaction Effects

```
!
! Kenny & Judd 1984 Psych Bull interaction example
!
#define nfac 6     ! number of factors including residuals
#define nfacsq 36 ! number of factors squared
#define nvar 4     ! number of observed variables
#define nvarsq 16 ! square of number of obs var
#define nindic 1   ! number of indicator variables, not squared
#define ntot 21    ! nvar + nvarsq + nindic
#define ninput 9   ! nvar*(nvar+3)/2 + nindic less any products not
                   ! used

Title Kenny & Judd interaction model in general form
 Data Ngroups=1 Ninput=ninput Nobs=500
 ! Simulated data from Kenny & Judd 1984 Psychol Bull p 205
 Labels X1 X2 Z1 Z2 X1Z1 X1Z2 X2Z1 X2Z2 Y
 CMatrix File=kennyint.cov

 Begin Matrices;
  B Full nvar nfac          ! Loadings on Linear factors (including
                            ! errors)
  S Symm nfac nfac          ! Covariances of Linear factors
  I Iden nfacsq nfacsq      ! Identity matrix
  J Full nfacsq nfacsq      ! Reordering matrix
  P Full nindic nfac        ! Paths from Linear factors to indicator
                            ! variables
  Q Full nindic nfacsq      ! Paths from Squared factors to
                            ! indicator variables
  E Diag nindic nindic Free ! Residual variance of indicator
                            ! variables
  F Full ninput ntot        ! Filter matrix to select some of the
                            ! variables
  Z Zero nvar nvarsq        ! Zeroes
 End Matrices;

 Specify B
  0 0 100 0 0 0
  101 0 0 102 0 0
  0 0 0 0 103 0
  0 104 0 0 0 105
 Matrix B
  1 0 .5 0 0 0
  .5 0 0 .5 0 0
```

```
 0 1 0 0 .5 0
 0 .5 0 0 0 .5
Labels row B X1 X2 Z1 Z2
Labels col B X Z ErrorX1 ErrorX2 ErrorZ1 ErrorZ2

Free S 1 1 S 2 1 S 2 2
Matrix S
 1
 0 1
 0 0 1
 0 0 0 1
 0 0 0 0 1
 0 0 0 0 0 1
Labels row S X Z ErrorX1 ErrorX2 ErrorZ1 ErrorZ2
Labels col S X Z ErrorX1 ErrorX2 ErrorZ1 ErrorZ2

Matrix E
 1
Labels row E YErrorVar
Labels col E YErrorVar

Matrix F
 1 0 0 0 0 0 0 0 0 0 0 0 0 0 0 0 0 0 0 0 0
 0 1 0 0 0 0 0 0 0 0 0 0 0 0 0 0 0 0 0 0 0
 0 0 1 0 0 0 0 0 0 0 0 0 0 0 0 0 0 0 0 0 0
 0 0 0 1 0 0 0 0 0 0 0 0 0 0 0 0 0 0 0 0 0
 0 0 0 0 0 0 1 0 0 0 0 0 0 0 0 0 0 0 0 0 0
 0 0 0 0 0 0 0 1 0 0 0 0 0 0 0 0 0 0 0 0 0
 0 0 0 0 0 0 0 0 0 0 1 0 0 0 0 0 0 0 0 0 0
 0 0 0 0 0 0 0 0 0 0 0 1 0 0 0 0 0 0 0 0 0
 0 0 0 0 0 0 0 0 0 0 0 0 0 0 0 0 0 0 0 0 1
Labels Row F
 X1 X2 Z1 Z2 X1Z1 X1Z2 X2Z1 X2Z2 Y
Labels Col F
 X1 X2 Z1 Z2
 X1X1 X1X2 X1Z1 X1Z1 X2X1 X2X2 X2Z1 X2Z1
 Z1X1 Z1X2 Z1Z1 Z1Z1 Z2X1 Z2X2 Z2Z1 Z2Z1 Y

Matrix J File=j6.mat

Free Q 1 2
Free P 1 1 P 1 2
Labels Row P Y
Labels Col P X Z ErrorX1 ErrorX2 ErrorZ1 ErrorZ2
Labels Row Q Y
Labels Col Q
```

```
  XX XZ XEX1 XEX2 XEZ1 XEZ2
  ZX ZZ ZEX1 ZEX2 ZEZ1 ZEZ2
  EX1X EX1Y EX1EX1 EX1EX2 EX1EZ1 EX1EZ2
  EX2X EX2Y EX2EX1 EX2EX2 EX2EZ1 EX2EZ2
  EZ1X EZ1Y EZ1EX1 EZ1EX2 EZ1EZ1 EZ1EZ2
  EZ2X EZ2Y EZ2EX1 EZ2EX2 EZ2EZ1 EZ2EZ2
 Begin Algebra;
  C = S@S*(I+J) ; ! Covariances among combinations
  M = B@B ;       ! Loadings on combinations
 End Algebra;

 Covariance F*( b*S*b' |     Z | b*S*p' _
                   Z' | m*C*m'| m*C*q' _
              p*S*b' | q*C*m'| p*S*p' + q*C*q' + E ) *F';
 Option GLS
End Group;
```

APPENDIX C
Mx Script for MacKenzie and Spreng Multigroup Example

```
!
! Interaction with multiple group approach
!
! Data from MacKenzie and Spreng (1992) How does motivation moderate
! the impact of central and peripheral processing on brand
! attitudes and intentions. Journal of Computer Research 18:519-529
!
! See also Rigdon, E et al (1998) A comparative review of interaction
! and nonlinear modeling In: R. Schumacker and G. Marcoulides (eds)
! Interaction and nonlinear effects in structural equation modeling.
! New York: Erlbaum
!
! define number of factors and number of observed variables

#define nvar 6
#define nfac 2

Title interaction for Ad & Brand attitude: Low group
 Data NInput=nvar Nobs=200 NGroups=2
 CMatrix Full File=low.cov
 Means File=low.mean
 Matrices
  A Full nfac 1           ! Factor Means
  D Diag nfac nfac Free   ! Factor Variances
  B Sdiag nfac nfac Free  ! Factor → Factor paths
  I Iden nfac nfac
  L Full nvar nfac        ! Factor → Observed paths
```

```
  E Diag nvar nvar Free  ! Residual Vars on Observed
  M Full nvar 1 Free     ! Score Means
 End Matrices

 Start 1 L 1 1 L 4 2
 Free L 2 1 L 3 1 L 5 2 L 6 2

 Mean L*(I-B)~*A + M ;
 Covariance L*(I-B)~*D*(I-B)~'*L'+ E ;

 Option RS
End Group
Title interaction for Ad & Brand attitude: High Group
 Data NInput=nvar Nobs=160 NGroups=2
 CMatrix Full File=high.cov
 Means File=high.mean
 Matrices = Group 1
  A Full nfac 1     Free ! Factor Means (different from group 1)
  B Sdiag nfac nfac Free ! Factor → Factor paths (different from
                           group 1)
 End Matrices

 Start 1 All

 Mean L*(I-B)~*A + M;
 Covariance L*(I-B)~*D*(I-B)~'*L'+ E ;

 Option Rsid
 Option Multiple
End Group

! write current results to binary file
  Save general.mxs

! Fit model with same slopes
  Equate B 1 2 1 B 2 2 1
  End
! Fit model with same means
 Get general.mxs
  Drop 20 21
  !Equate A 2 1 1 A 1 1 1
  !Equate A 2 2 1 A 1 2 1
  End
! Fit model with same means and same slopes
  Equate B 1 2 1 B 2 2 1
End
```

4

EQS and LISREL Examples Using Survey Data

Robert A. Ping, Jr.
Wright State University

In studies reported in the social science literature that involve categorical independent variables (analysis of variance [ANOVA] studies), interactions (e.g., XZ in $Y = b_0 + b_1X + b_2Z + b_3XZ + b_4XX$) and quadratics ($XX$ in the equation just mentioned) are routinely investigated to enable the interpretation of significant first-order (main) effects (i.e., b_1 and b_2). Interaction and quadratic variables are also investigated in studies involving continuous variables and regression, although not routinely, and not to aid in the interpretation of significant first-order effects as they are in ANOVA studies. Typically, continuous interactions and quadratics are investigated in response to theory that proposes their existence.

Researchers in the social sciences have called for the estimation of continuous interaction and quadratic effects to improve the interpretation of significant first-order effects (Aiken & West, 1991; Blalock, 1965; J. Cohen, 1968; J. Cohen & P. Cohen, 1975, 1983; Howard, 1989; Jaccard, Turrisi, & Wan, 1990; Kenny, 1985). Their argument is that failing to consider the presence of interactions and quadratics in the population model increases the risk of false negative research findings, and positive research findings that are conditional in the population equation. To explain, in a model such as $Y = b_0 + b_1X + b_2Z + b_3XZ$, the coefficient of Z is $(b_2 + b_3X)$. The statistical significance of this coefficient could be very different from the statistical significance of the Z coefficient in the same model specified without the XZ interaction (i.e., $Y = b_0' + b_1'X + b_2'Z$). Specifically, b_2' could be nonsignificant whereas $b_2 + b_3X$ could be significant over part(s) of the range of X. Failing to specify this significant interaction would lead to a false disconfirmation of the Z–Y association. Alternatively b_2' could be significant whereas $b_2 + b_3X$ could be nonsignificant over part of the range of X. Failing to specify this significant interaction could produce a misleading picture of the contingent Z–Y association. The implications for failing to specify a population quadratic are similar.

Fortunately there has been considerable progress in the estimation of interactions and quadratics using structural equation analysis recently. Several specification techniques have been proposed (see Hayduk, 1987; Kenny & Judd, 1984; Ping, 1995, 1996a; Wong & Long, 1987), and LISREL 8 has become commercially available. Along with subgroup analysis (Jöreskog, 1971a), these alternatives offer substantive researchers considerable power and flexibility in estimating latent variable interaction and quadratic effects in structural equation models. After briefly summarizing these alternatives, illustrations using field survey data, EQS and LISREL 8 are provided. The chapter concludes with suggested procedures for these several techniques, and a discussion of topics for further research in this area.

LATENT VARIABLE INTERACTION AND QUADRATIC SPECIFICATION AND ESTIMATION

Estimation techniques for interaction and quadratic latent variables in structural equation models can be classified into direct and indirect approaches. Direct estimation approaches produce structural coefficient estimates without introducing additional convenience variables to the model. Examples include multiple-indicator specification (Kenny & Judd, 1984; see Jaccard & Wan, 1990; Jöreskog & Yang, 1996) and single-indicator specification (see Ping, 1995). These specifications can be estimated directly with COSAN, LISREL 8, and similar software that accommodates nonlinear constraint equations.[1]

With indirect estimation approaches structural coefficient estimates may or may not be available, and convenience variables or several estimation steps are required. Examples include subgroup analysis (Jöreskog, 1971a), convenience-variable techniques (Hayduk, 1987; Wong & Long, 1987), indirect multiple-indicator specification (Ping, 1996a), and indirect single-indicator specification (Ping, 1995). These techniques can be used with EQS and the other structural equation software packages mentioned earlier.

---

[1]COSAN is available from SAS, Inc. Other software that will directly estimate latent variable interactions and quadratics includes LINCS (distributed by APTEC Systems), RAMONA (distributed by Michael W. Browne, Ohio State University), and MECOSA (distributed by SLI-AG, Frauenfeld, Switzerland).

Direct Estimation

Kenny and Judd (1984) proposed that latent variable interactions and quadratics could be specified adequately using all unique products of the indicators of their constituent linear latent variable. For the linear latent variables X and Z with indicators x_1, x_2, z_1, z_2, and z_3, respectively, the latent variable interaction XZ can be specified with the product indicators x_1z_1, x_1z_2, x_1z_3, x_2z_1, x_2z_2, and x_2z_3. X^*X can be specified with the product indicators x_1x_1, x_1x_2, and x_2x_2. Under the Kenny and Judd normality assumptions,[2] the variance of the product indicator x_1z_1, for example, is given by

$$\begin{aligned} \mathrm{Var}(x_1z_1) &= \mathrm{Var}[(\lambda_{x1}X + \varepsilon_{x1})(\lambda_{z1}Z + \varepsilon_{z1})] \\ &= \lambda_{x1}{}^2\lambda_{z1}{}^2\mathrm{Var}(XZ) + \lambda_{x1}{}^2\mathrm{Var}(X)\mathrm{Var}(\varepsilon_{z1}) \\ &\quad + \lambda_{z1}{}^2\mathrm{Var}(Z)\mathrm{Var}(\varepsilon_{x1}) + \mathrm{Var}(\varepsilon_{x1})\mathrm{Var}(\varepsilon_{z1}), \end{aligned} \tag{1}$$

where λ_{x1} and λ_{z1} are the loadings of the indicators x_1 and z_1 on the latent variables X and Z; ε_{x1} and ε_{z1} are the error terms for x_1 and z_1; Var(a) is the variance of a; and Cov(X,Z) is the covariance of X and Z. The variance of XZ is given by

$$\mathrm{Var}(XZ) = \mathrm{Var}(X)\mathrm{Var}(Z) + \mathrm{Cov}(X,Z)^2 \tag{1a}$$

(Kendall & Stuart, 1958).

In the quadratic case the variance of the product indicator x_1x_1 is given by

$$\begin{aligned} \mathrm{Var}(x_1x_1) &= \mathrm{Var}[(\lambda_{x1}X + \varepsilon_{x1})^2] \\ &= \mathrm{Var}[\lambda_{x1}{}^2X^2 + 2\lambda_{x1}{}^2\varepsilon_{x1} + \varepsilon_{x1}{}^2] \\ &= \lambda_{x1}{}^2\lambda_{x1}{}^2\mathrm{Var}(X^2) + 4\lambda_{x1}{}^2\mathrm{Var}(X)\mathrm{Var}(\varepsilon_{x1}) + 2\mathrm{Var}(\varepsilon_{x1})^2. \end{aligned} \tag{2}$$

The variance of XX is given by

$$\mathrm{Var}(XX) = 2\mathrm{Var}(X)^2 \tag{2a}$$

using the Equation 1a result.

Kenny and Judd (1984) provided examples of a COSAN implementation of their technique, and examples of a LISREL 8 implementation of their

[2]The Kenny and Judd (1984) normality assumptions were that each of the latent variables X and Z is independent of the errors (ε_{x1}, ε_{x2}, ε_{z1}, and ε_{z2}), the error terms are mutually independent, the indicators x_1, x_2, z_1, and z_2 are multivariate normal, and the errors (ε_{x1}, ε_{x2}, ε_{z1}, and ε_{z2}) are multivariate normal.

technique are provided in Jaccard and Wan (1995) and Jöreskog and Yang (1996). The implementation procedure involves centering the indicators for X, Z and Y;[3,4] adding the product indicators for XZ and/or XX to each case; specifying XZ and/or XX in the structural model with the appropriate product indicators, constraining the loadings and error variances of these product indicators to equal their Equation 1 and 2 forms, constraining the variance of XZ and XX to their Equation 1a and 2a forms, and subsequently estimating this model with COSAN, LISREL 8, or other software that accommodates nonlinear constraints. The loading of the product indicator x_1z_1, for instance, would be constrained to equal $\lambda_{x1}\lambda_{z1}$ in Equation 1, and the error variance of x_1z_1 would be constrained to equal $\lambda_{x1}{}^2\text{Var}(X)\text{Var}(\varepsilon_{z1}) + \lambda_{z1}{}^2\text{Var}(Z)\text{Var}(\varepsilon_{x1}) + \text{Var}(\varepsilon_{x1})\text{Var}(\varepsilon_{z1})$, also in Equation 1. Similarly, the loading of x_1x_1, for example, would be constrained to equal $\lambda_{x1}\lambda_{x1}$ in Equation 2, and the error variance of x_1x_1 would be constrained to equal $4\lambda_{x1}{}^2\text{Var}(X)\text{Var}(\varepsilon_{x1}) + 2\text{Var}(\varepsilon_{x1})^2$, also in Equation 2.

This technique is powerful because it models several interaction and/or quadratic latent variables and provides coefficient estimates for these variables, but it has been infrequently used in the substantive literature. This may have been because it is not widely known, and can be difficult to use (Aiken & West, 1991; see Jaccard & Wan, 1995; Jöreskog & Yang, 1996). Coding the constraint equations for the product indicators in COSAN can be a daunting task. LISREL 8 provides a nonlinear constraint capability that is different from that available in COSAN, and the constraint coding effort is reduced. However, this task can still become tedious for larger models (Jöreskog & Yang, 1996; Ping, 1995). In addition, for larger models, the size of the covariance matrix created by the addition of product indicators, and the number of additional variables implied by the constraint equations, can create model convergence and other model estimation problems.

As an alternative to the Kenny and Judd (1984) product-indicator specification, single-indicator specification has been proposed (Ping, 1995). The product-indicators of XZ and XX are replaced by single indicators $x{:}z = [(x_1 + x_2)/2][z_1 + z_2 + z_3)/3]$ for XZ and $x{:}x = [(x_1 + x_2)/2][(x_1 + x_2)/2]$ for XX. The

---

[3]Centering an indicator involves subtracting the mean of the indicator from each case value. As a result the indicator has a mean of zero (see Aiken & West, 1991; Bollen, 1989; Cronbach, 1987; Jaccard et al., 1990; Kenny & Judd, 1984). Although this is a standard assumption in structural equation modeling for variables with an arbitrary zero point such as Likert-scaled and other rating-scaled variables (see Bollen, 1989), mean or zero centering has been the subject of much confusion over the interpretation of centered variables. Aiken and West presented a compelling and exhaustive argument for the efficacy of centering.

[4]Centering the indicators of the dependent variable Y is optional and has no effect on the structural coefficient estimates. However, it is recommended and used throughout this presentation to produce coefficient estimates that are equivalent to those produced when an intercept is specified (see Footnote 12).

loadings and error variances for these single indicators are similar in appearance to Equations 1 and 2:

$$\begin{aligned}
\mathrm{Var}(x{:}z) &= \mathrm{Var}[(x_1 + x_2)/2][(z_1 + z_2 + z_3)/3] \\
&= \mathrm{Var}\{(x_1 + x_2)/2\}\mathrm{Var}\{(z_1 + z_2 + z_3)/3\} + \mathrm{Cov}\{(x_1 + x_2)/2,(z_1 + z_2 + z_3)/3\}^2 \\
&= [\Gamma_X{}^2\mathrm{Var}(X) + \theta_X][\Gamma_Z{}^2\mathrm{VAR}(Z) + \theta_Z] + [\Gamma_X\Gamma_Z\mathrm{Cov}(X,Z)]^2 \\
&= \Gamma_X{}^2\Gamma_Z{}^2[\mathrm{Var}(X)\mathrm{Var}(Z) + \mathrm{Cov}(X,Z)^2] + \Gamma_X{}^2\mathrm{Var}(X)\theta_Z \\
&\quad + \Gamma_Z{}^2\mathrm{Var}(Z)\theta_X + \theta_X\theta_Z \\
&= \Gamma_X{}^2\Gamma_Z{}^2\mathrm{Var}(XZ) + \Gamma_X{}^2\mathrm{Var}(X)\theta_Z + \Gamma_Z{}^2\mathrm{Var}(Z)\theta_X + \theta_X\theta_Z \qquad (3)
\end{aligned}$$

where $\Gamma_X = (\lambda_{x1} + \lambda_{x2})/2$, $\theta_X = (\mathrm{Var}(\varepsilon_{x1}) + \mathrm{Var}(\varepsilon_{x2}))/2^2$, $\Gamma_Z = (\lambda_{z1} + \lambda_{z2} + \lambda_{z3})/3$, $\theta_Z = (\mathrm{Var}(\varepsilon_{z1}) + \mathrm{Var}(\varepsilon_{z2}) + \mathrm{Var}(\varepsilon_{z3}))/3^2$.

Similarly,

$$\begin{aligned}
\mathrm{Var}(x{:}x) &= \mathrm{Var}\{[(x_1 + x_2)/2][(x_1 + x_2)/2]\} \\
&= 2\mathrm{Var}\{(x_1 + x_2)/2\}^2 \\
&= 2\mathrm{Var}[(\lambda_{x1}X + \varepsilon_{x1}) + (\lambda_{x2}X + \varepsilon_{x2})]^2/2^4 \\
&= 2[\mathrm{Var}((\lambda_{x1} + \lambda_{x2})X + \mathrm{Var}(\varepsilon_{x1}) + \mathrm{Var}(\varepsilon_{x2})]^2/2^4 \\
&= 2[\mathrm{Var}(\Gamma_X X) + \theta_X]^2/2^4 \\
&= \Gamma_X{}^4\mathrm{Var}(X^2) + 4\Gamma_X{}^2\mathrm{Var}(X)\theta_X + 2\theta_X{}^2. \qquad (4)
\end{aligned}$$

These results extend to latent variables with an arbitrary number of indicators.[5]

Estimation using single-indicator specification also involves centering the indicators for *X, Z,* and *Y* (see Footnotes 4 and 12); adding the single indicators *x:z* and/or *x:x* to each case; specifying *XZ* and/or *XX* in the structural model using the single indicators *x:z* and/or *x:x,* constraining the loadings and error variances of these single indicators to equal their Equation 3 and 4 forms, and constraining the variance of *XZ* and/or *XX* to their Equation 1a and 2a forms; and subsequently estimating this structural model using COSAN, LISREL 8, or similar software that provides a nonlinear constraint capability. The loading of the single indicator *x:z,* for instance, would be constrained to equal $\Gamma_X\Gamma_Z$ in Equation 3, and the error variance of *x:z* would be constrained to equal $\Gamma_X{}^2\mathrm{Var}(X)\theta_Z + \Gamma_Z{}^2\mathrm{Var}(Z)\theta_X + \theta_X\theta_Z$, also in Equation 3. Similarly, the loading of *x:x,* for example, would be constrained to equal $\Gamma_X\Gamma_X$ in Equation 4, and the error variance of *x:x* would

---

[5]By induction, for an arbitrary latent variable X, $\Gamma_X = (\lambda_{x1} + \lambda_{x2} + \ldots + \lambda_{zm})/m$ and $\theta_X = (\mathrm{Var}(\varepsilon_{x1}) + \mathrm{Var}(\varepsilon_{x2}) + \ldots + \mathrm{Var}(\varepsilon_{xm}))/m^2$, where m is the number of indicators of *X*, and Equations 3 and 4 apply to *X* and *Z* with an arbitrary number of indicators.

be constrained to equal $4\Gamma_X^2 Var(X)\theta_X + 2\theta_X^2$, also in Equation 4. An example using this technique is provided later in the chapter.

Single-indicator specification appears to be equivalent to the Kenny and Judd (1984) specification (Ping, 1995); it has the power of the Kenny and Judd technique because it can model several interaction and/or quadratic variables and provide coefficient estimates for these variables. In addition, it requires less specification effort than the Kenny and Judd approach, and produces an input covariance matrix with fewer elements.[6] However, it is new and has yet to appear in the substantive literature.

Indirect Estimation

Subgroup Analysis. Subgroup analysis (Jöreskog, 1971a) generally involves splitting the sample and assessing differences in model fit when the model is restricted to the resulting groups of cases. The procedure involves dividing a sample into subgroups of cases based on different levels (e.g., low and high) of a suspected interaction *X*. Constraining the model coefficients to be equal between subgroups for the model estimated in each subgroup, the coefficients of the linear-terms-only model (e.g., the model without any interaction or quadratic latent variables present) are then estimated for this model using each of the resulting subgroups and structural equation analysis. The result is a chi-square statistic for the model's fit across the two subgroups with this coefficient equality constraint between the subgroups. Relaxing this equality constraint, the model is re-estimated, and the resulting chi-square statistic is compared with that from the first estimation. A significant difference between the chi-square statistics for these two nested models suggests that there is at least one coefficient difference between the two groups. The coefficients from the second estimation are then tested for significant differences between the groups using a coefficient difference test (see, e.g., Jaccard et al., 1990). A significant coefficient difference between the *Z* coefficients suggests an interaction between that variable and *X*, the variable used to create the subgroups. A significant coefficient difference between the *X* coefficients suggests the presence of a quadratic.

This technique is popular in the substantive literature, and is a preferred technique in some situations. Jaccard et al. (1990) stated that subgroup analysis may be appropriate when the model could be structurally different for different subgroups of subjects. They also pointed out that an interaction

[6]Each additional product indicator adds an input variable, and a row and column, to the sample covariance matrix. Adding product indicators can become statistically detrimental in that each additional product indicator places additional demands on the sample covariance matrix: The number of resulting variables can become too large to yield a reasonably stable matrix (Jaccard & Wan, 1995).

need not be of the form "X times Z" (interaction forms include X/Z, and the possibilities are infinite; Jaccard et al., 1990; see also Hanushek & Jackson, 1977), and that three-group analysis may be more appropriate in these cases. Sharma, Durand, and Gur-Arie (1981) recommended subgroup analysis to detect what they termed a homologizer: a variable W, for example, that affects the strength of an association between two variables, X and Y, yet is not related to either X or Y. However, subgroup analysis is criticized in the regression literature for its reduction of statistical power and increased likelihood of Type II error (J. Cohen & P. Cohen, 1983; Jaccard et al., 1990; see also Bagozzi, 1992). In addition, coefficient estimates for significant interactions or quadratics are not available in subgroup analysis, and interpretation of a significant interaction or quadratic is problematic. Sample size requirements also limit the utility of subgroup analysis. Samples of 100 cases per subgroup are considered by many to be the minimum sample size, and 200 cases per group are usually recommended (Boomsma, 1983; see Gerbing & J. C. Anderson, 1985, for an alternative view).

Convenience Variables. Hayduk (1987) and Wong and Long (1987) suggested an approach involving the addition of convenience variables to the structural model in order to accomplish the estimation of an interaction or quadratic effect. Hayduk, for instance, suggested specifying the loading of x_1z_1 on XZ by creating an intervening latent variable between XZ and x_1z_1 that has an error of zero. The indirect effect of XZ on x_1z_1 via this variable could then be used to specify the loading of x_1z_1 on XZ. The error term for x_1z_1 is handled similarly.

These convenience variable approaches are powerful because they can specify multiple interaction and quadratic variables, and provide coefficient estimates for interactions and quadratics. However the effort required to specify a model using these techniques can be considerable (see Hayduk, 1987). Perhaps as a result they are infrequently used in the substantive literature.

Two-Step Approaches. Two-step estimation techniques involve calculating the loadings and error variance for the (product or single) indicator(s) of XZ using measurement model parameter estimates, then fixing these loadings and error variances at their calculated values in the structural model.

Estimates of the parameters comprising the loadings and error variances of the indicator(s) of latent variable interactions and quadratics are available in a linear-terms-only measurement model corresponding to the structural model of interest (i.e., a measurement model that includes the linear latent variables but excludes the interactions and quadratics). With sufficient unidimensionality, that is the indicators of each construct have only one underlying construct each (Aaker & Bagozzi, 1979; J. C. Anderson & Gerbing,

1988; Burt, 1973; Hattie, 1985; Jöreskog, 1970, 1971b; McDonald, 1981), these measurement model parameter estimates change trivially, if at all, between the measurement model and alternative structural models (J. C. Anderson & Gerbing, 1988). As a result, instead of specifying the loadings and error variances for the indicators of *XZ* or *XX* as variables in the structural model, they can be calculated using parameter estimates from the linear-terms-only measurement model (i.e., involving only *X, Z,* and *Y*), and specified as constants in the structural model if *X* and *Z* are each sufficiently unidimensional. This is possible because the unidimensionality of *X* and *Z* permits the omission of the *XZ* and *XX* latent variables from the linear-terms-only measurement model: Because *X* and *Z* are each unidimensional, their indicator loadings and error variances are unaffected by the presence or absence of other latent variables in a measurement or structural model, in particular *XZ* and *XX*.

To use this technique *X, Z,* and *Y* are unidimensionalized,[7] *X, Z,* and Y are centered, and the (product or single) indicator(s) for *XZ* and/or *XX* are added to each case as before. Then the Equation 1, 1a, 2, 2a, 3, and/or 4 parameters (i.e., λ_{x1}, λ_{x2}, λ_{z1}, λ_{z2}, λ_{z3}, Var(ε_{x1}), Var(ε_{x2}), Var(ε_{z1}), Var(ε_{z2}), Var(ε_{z3}), Var(*X*), Var(*Z*), and Cov(*X,Z*)) are estimated in a linear-terms-only measurement model (e.g., one that excludes *XX* and *XZ*). Then, the loadings and error variances for the indicators of *XZ* and/or *XX,* plus the variances of *XZ* and/or *XX* are calculated by substituting these measurement model parameter estimates into the appropriate versions of Equations 1 through 4.[8] Finally, using a structural model in which these calculated loadings and error variances for the indicators of *XZ* and/or *XX,* plus the variance of *XZ* and/or *XX,* are specified as constants (i.e., fixed rather than free), the structural model is estimated using EQS or LISREL.[9] An example of this technique is provided in the next section.

If the structural model estimates of the linear latent variable measurement parameters (e.g., λ_{x1}, λ_{x2}, λ_{z1}, λ_{z2}, λ_{z3}, Var(ε_{x1}), Var(ε_{x2}), Var(ε_{z1}), Var(ε_{z2}), Var(ε_{z3}), Var(*X*), Var(*Z*), and Cov(*X,Z*)) are not similar to the linear-terms-only measurement model estimates (i.e., equal in the first two decimal places) and the calculated values fixed in the specification of *XZ* and/or *XX* in the structural model therefore do not adequately reflect the structural model

[7]Procedures for obtaining unidimensionality are suggested in J. C. Anderson and Gerbing (1982), Gerbing and J. C. Anderson (1988), and Jöreskog (1993).

[8]Ping (1996a) showed that in general the loading of a product indicator *xz* is given by $\lambda_x\lambda_z$ and the error variance is given by $K\lambda_x^2$Var(*X*)Var(ε_z) + $K\lambda_z^2$Var(*Z*)Var(ε_x) + *K*Var(εx)Var(εz) (*K* = 2 if *x* = *z*, *K* = 1 otherwise). For the single-indicator technique, λ and ε in this equation are replaced by Γ and θ defined at Equation 3.

[9]LISREL 8 is available for microcomputers only, and according to individuals at SSI when this chapter was written, there are no plans to release a mainframe version of LISREL 8 in the future. As a result mainframe LISREL 7 is likely to remain in use. The two-step technique can also be used with LISREL 8 in problem situations.

measurement parameter values, the Equation 1 and 1a, 2 and 2a, 3, or 4 values can be recomputed using the structural model estimates of these measurement parameters. The structural model can then be re-estimated with these updated fixed values. Zero to two of these re-estimations are usually sufficient to produce consecutive structural models with the desired trivial difference in measurement parameters and coefficient estimates that are equivalent to direct LISREL 8 or COSAN estimates.

Two-step techniques appear to be equivalent to direct estimation (Ping, 1995, 1996a). They have the power of the Kenny and Judd (1984) technique because they can model several interaction and/or quadratic variables and provide coefficient estimates for these variables. In addition they require less specification effort than the direct approaches, and with a single indicator, produce a sample covariance matrix with fewer elements than the Kenny and Judd approach. However, two-step techniques are also new.

EXAMPLES

In the marketing literature, Ping (1993) reported that relationship neglect (reduced physical contact, NEG) in a buyer–seller relationship (i.e., an economic plus social exchange relationship) had antecedents that included overall relationship satisfaction (SAT), alternative attractiveness (ALT), and relationship investment (INV) (see also Rusbult, Zembrodt, & Gunn, 1982). However, a hypothesized switching cost (SCT) association with NEG was not significant.

Recalling the previous remarks concerning the role of population interactions or quadratics in disconfirmed theory test results, the nonsignificant SCT-NEG effect may have been due to an unmodeled interaction or quadratic present in the population model. Because a quadratic SCT-NEG effect and several SCT interaction effects were plausible, the SCT*SCT quadratic, and the SCT interactions with the other antecedents were estimated using direct estimation, a single indicator per interaction or quadratic latent variable, and the following structural model,

$$\begin{aligned} NEG = \beta_1 SAT + \beta_2 ALT + \beta_3 INV + \beta_4 SCT + \beta_5 SAT*SCT \\ + \beta_6 ALT*SCT + \beta_7 INV*SCT + \beta_8 SCT*SCT + \zeta. \end{aligned} \quad (5)$$

The LISREL 8 program code for this specification is shown in Fig. 4.1, the resulting covariance matrix for the indicators is shown in Table 4.1, and the results are shown in Table 4.2.

The interactions of SCT with SAT and INV were significant when all the interactions and the quadratic latent variable involving SCT were jointly specified (see Table 4.1). When these interactions and the quadratic were estimated individually, however, only the INV*SCT interaction was signifi-

```
NEGLECT with nonlinears (Full Structural Model)
DA NI=53 NO=222
LA
 sa1 sa2 sa3 sa4 sa5 sa6 sa7 al1 al2 al3 al4 al5 al6
 in1 in2 in3 in4 in5 in6 sc1 sc2 sc3 sc4 sc5
 ne1 ne2 ne3 ne4 ne5 ne6 ne7
 x1 x2 x3 x4 x5 x6 x7 x8 x9 x10
 ssc x12 x13 asc x14 isc scsc x15 x16 x17 x18 x19
RA FI=d:\lisrel8w\neg\neg1.dat FO
(f9.6,7f10.6/f9.6,7f10.6/f9.6,7f10.6/f9.0,6f10.0,f10.4/f9.4,7f10.4/f9.4,
 7f10.6/f9.6,4f10.6)
SE
 ne2 ne5 ne6 ne7
 sa2 sa4 sa5 sa6 sa7 al2 al3 al4 al5 in1 in3 in4 in5 sc2 sc3 sc4 sc5
 ssc asc isc scsc
/
MO NY=4 NX=21 ne=1 nk=8 ap=8
LE
NEG
LK
SAT ALT INV SWC
SSC ASC ISC SCSC
pa ly
*
1
1
1
0
ma ly
*
 .69415
 .73074
 .87250
1.00000
pa te
*
1 1 1 1
ma te
*
 .19060 .10722 .06301 .14102
pa lx
*
1 0 0 0 0 0 0 0
1 0 0 0 0 0 0 0
1 0 0 0 0 0 0 0
1 0 0 0 0 0 0 0
1 0 0 0 0 0 0 0
0 1 0 0 0 0 0 0
0 1 0 0 0 0 0 0
0 1 0 0 0 0 0 0
0 1 0 0 0 0 0 0
0 0 1 0 0 0 0 0
```

FIG. 4.1. *(Continued)*

```
0 0 1 0 0 0 0 0
0 0 1 0 0 0 0 0
0 0 1 0 0 0 0 0
0 0 0 1 0 0 0 0
0 0 0 1 0 0 0 0
0 0 0 1 0 0 0 0
0 0 0 1 0 0 0 0
0 0 0 0 1 0 0 0
0 0 0 0 0 1 0 0
0 0 0 0 0 0 1 0
0 0 0 0 0 0 0 1
ma lx
*
 .57047  0     0     0     0 0 0 0
 .63797  0     0     0     0 0 0 0
 .71961  0     0     0     0 0 0 0
 .63319  0     0     0     0 0 0 0
 .67682  0     0     0     0     0     0     0
     0 .85382  0     0     0     0     0     0
     0 .83309  0     0     0     0     0     0
     0 .92148  0     0     0     0     0     0
     0 .72202  0     0     0     0     0     0
     0     0 .69181  0     0     0     0     0
     0     0 .76535  0     0     0     0     0
     0     0 .77603  0     0     0     0     0
     0     0 .75915  0     0     0     0     0
     0     0     0 .88871  0     0     0     0
     0     0     0 .96722  0     0     0     0
     0     0     0 .95239  0     0     0     0
     0     0     0 .98422  0     0     0     0
     0     0     0     0 .61402  0     0     0
     0     0     0     0     0 .79433  0     0
     0     0     0     0     0     0 .70928  0
     0     0     0     0     0     0     0 .89896
pa td
*
21*1
ma td
*
 .16702 .13028 .10938 .11966 .10219
 .27195 .24934 .07797 .24416
 .45436 .12037 .08948 .12429
 .29309 .21006 .17536 .20934
 .04726 .03345 .05733 .20569
pa ga
*
!sa al in sc ssc asc isc scsc
  1  1  1  1  1   1   1   1 !ne
pa ph
```

FIG. 4.1. *(Continued)*

```
*
  0 !sa
  1  0 !al
  1  1  0 !in
  1  1  1  0 !sc
  0  0  0  0  0 !ssc
  0  0  0  0  0   0 !asc
  0  0  0  0  0   0   0 !isc
  0  0  0  0  0   0   0   0 !scsc
!sa al in sc ssc asc isc scsc
ma ph
*
  1 !sa
-.55972    1 !al
 .35660 -.28679   1 !in
 .26639 -.40420 .53819 1 !sc
   0       0       0    0  1 !ssc
   0       0       0    0  0   1 !asc
   0       0       0    0  0   0   1 !isc
   0       0       0    0  0   0   0   1 !scsc
! sa      al      in   sc ssc asc isc scsc
!par(1)=lsat, par(2)=lsct, par(3)=lalt, par(4)=linv
co par(1)=.2*lx(1,1)+.2*lx(2,1)+.2*lx(3,1)+.2*lx(4,1)+.2*lx(5,1)
co par(2)=.25*lx(14,4)+.25*lx(15,4)+.25*lx(16,4)+.25*lx(17,4)
co lx(18,5)=par(1)*par(2)
co par(3)=.25*lx(6,2)+.25*lx(7,2)+.25*lx(8,2)+.25*lx(9,2)
co lx(19,6)=par(3)*par(2)
co par(4)=.25*lx(10,3)+.25*lx(11,3)+.25*lx(12,3)+.25*lx(13,3)
co lx(20,7)=par(4)*par(2)
co lx(21,8)=par(2)*par(2)
!par(1)=lsat, par(2)=lsct, par(3)=lalt, par(4)=linv
!par(5)=tsat, par(6)=tsct, par(7)=talt, par(8)=tinv
co par(5)=.04*td(1,1)+.04*td(2,2)+.04*td(3,3)+.04*td(4,4)+.04*td(5,5)
co par(6)=.0625*td(14,14)+.0625*td(15,15)+.0625*td(16,16)+.0625*td(17,17)
!                  psat*tsct*lsat^2 + psct*tsat*lsct^2 + tsat*tsct
co td(18,18)=ph(1,1)*par(6)*par(1)^2+ph(4,4)*par(5)*par(2)^2+par(5)*par(6)
!                  4psct*tsct*lsct^2 + 2*tsct*tsct
co par(7)=.0625*td(6,6)+.0625*td(7,7)+.0625*td(8,8)+.0625*td(9,9)
!                  palt*tsct*lalt^2 + psct*talt*lsct^2 + talt*tsct
co td(19,19)=ph(2,2)*par(6)*par(3)^2+ph(4,4)*par(7)*par(2)^2+par(7)*par(6)
co par(8)=.0625*td(10,10)+.0625*td(11,11)+.0625*td(12,12)+.0625*td(13,13)
!                  pinv*tsct*linv^2 + psct*tinv*lsct^2 + tinv*tsct
co td(20,20)=ph(3,3)*par(6)*par(4)^2+ph(4,4)*par(8)*par(2)^2+par(8)*par(6)
co td(21,21)=4*ph(4,4)*par(6)*par(2)^2+2*par(6)^2
OU xm nd=5 it=100 ad=off
```

FIG. 4.1. Single-Indicator LISREL8 Specification Code

TABLE 4.1
SAT, ALT, INV, SCT, SAT:SCT, ALT:SCT, INV:SCT, SCT:SCT COVARIANCES[a]

	ne2	*ne5*	*ne6*	*ne7*	*sa2*	*sa4*
ne2	0.44140					
ne5	0.27223	0.38517				
ne6	0.31352	0.33582	0.45926			
ne7	0.35339	0.37251	0.45510	0.66153		
sa2	–0.15988	–0.15611	–0.16591	–0.23831	0.49246	
sa4	–0.18438	–0.18466	–0.22824	–0.33708	0.36048	0.53728
sa5	–0.18536	–0.16956	–0.22200	–0.33904	0.41704	0.46072
sa6	–0.18047	–0.17042	–0.20342	–0.27496	0.37179	0.40007
sa7	–0.17651	–0.17989	–0.23183	–0.31230	0.37707	0.43007
al2	0.25955	0.21616	0.26864	0.33358	–0.28884	–0.34299
al3	0.20668	0.17537	0.20619	0.29571	–0.25339	–0.30035
al4	0.25592	0.21004	0.26481	0.31275	–0.26772	–0.33390
al5	0.20333	0.12988	0.18715	0.27997	–0.21569	–0.29722
in1	–0.12763	–0.12174	–0.13297	–0.18287	0.11388	0.16758
in3	–0.15336	–0.15800	–0.15018	–0.22074	0.14480	0.18307
in4	–0.15580	–0.13721	–0.15042	–0.21658	0.18552	0.20435
in5	–0.13876	–0.12896	–0.12821	–0.18250	0.13650	0.18250
sc2	–0.13277	–0.12333	–0.18055	–0.23387	0.17270	0.23851
sc3	–0.11039	–0.11447	–0.15523	–0.17101	0.09804	0.21259
sc4	–0.12079	–0.09851	–0.14043	–0.19180	0.12745	0.20574
sc5	–0.13852	–0.12199	–0.16709	–0.18201	0.13198	0.17676
ssc	0.07535	0.04210	0.05847	0.05724	–0.01379	–0.04734
asc	–0.02316	–0.00725	–0.02152	–0.00257	0.01542	0.07090
isc	–0.06578	–0.05693	–0.05413	–0.04888	0.02676	0.02959
scsc	–0.08147	–0.08957	–0.07690	–0.11190	0.09952	0.07296

	sa5	*sa6*	*sa7*	*al2*	*al3*	*al4*
sa5	0.62723					
sa6	0.44855	0.52059				
sa7	0.48742	0.43594	0.56027			
al2	–0.35895	–0.29047	–0.31580	1.00096		
al3	–0.32604	–0.24288	–0.31442	0.71249	0.94338	
al4	–0.38545	–0.32002	–0.34312	0.79029	0.76544	0.92709
al5	–0.32559	–0.22898	–0.29530	0.59027	0.61808	0.66606
in1	0.17229	0.10124	0.13057	–0.27482	–0.17733	–0.19259
in3	0.20973	0.17598	0.17720	–0.25107	–0.14101	–0.16767
in4	0.24251	0.17843	0.19164	–0.28531	–0.19164	–0.22527
in5	0.17572	0.13047	0.12971	–0.23152	–0.14932	–0.18024
sc2	0.24383	0.17425	0.20884	–0.40555	–0.35062	–0.37555
sc3	0.17969	0.11793	0.12250	–0.40218	–0.28238	–0.34601
sc4	0.21265	0.15774	0.16347	–0.34183	–0.33240	–0.33629
sc5	0.16972	0.12117	0.12103	–0.34575	–0.28091	–0.33421
ssc	–0.06763	–0.05984	–0.03617	0.07300	0.01224	0.05605
asc	0.04521	0.07436	0.04472	0.00805	0.09021	–0.00207
isc	0.04741	0.05515	0.03217	–0.01035	–0.05486	–0.04765
scsc	0.14823	0.10550	0.13885	–0.06222	–0.10941	–0.06818

(Continued)

TABLE 4.1
(Continued)

	al5	*in1*	*in3*	*in4*	*in5*	*sc2*
al5	0.76548					
in1	−0.13126	0.93296				
in3	−0.10799	0.53724	0.70613			
in4	−0.14724	0.53430	0.59178	0.69170		
in5	−0.11463	0.50151	0.58371	0.59276	0.70060	
sc2	−0.33452	0.52046	0.38535	0.39525	0.35671	1.08290
sc3	−0.25564	0.56859	0.39428	0.40932	0.44495	0.84950
sc4	−0.31727	0.49955	0.35514	0.37129	0.37481	0.85200
sc5	−0.26591	0.55972	0.35306	0.35698	0.38386	0.86839
ssc	0.05921	0.07227	0.00451	0.02310	0.05299	0.09869
asc	0.05148	−0.02698	−0.03816	−0.03706	−0.04185	−0.07080
isc	−0.03119	−0.13881	−0.12571	−0.12979	−0.13205	−0.02828
scsc	−0.06057	−0.09622	0.02039	0.01089	−0.00445	−0.06396

	sc3	*sc4*	*sc5*	*ssc*	*asc*	*isc*
sc3	1.14557					
sc4	0.91651	1.08241				
sc5	0.95883	0.94152	1.17804			
ssc	0.10792	0.11839	0.12704	0.42458		
asc	−0.04004	−0.09487	−0.09467	−0.28824	0.69995	
isc	−0.06000	0.00575	0.01315	0.14864	−0.21973	0.58709
scsc	−0.11561	−0.00370	−0.03660	0.15748	−0.39768	0.41997

	scsc
scsc	0.94094

[a]The Table 4.2 and 4.3 results were obtained from raw data. Estimates using these covariances will be slightly different.

cant. The results for Equation 5 respecified with only the significant INV*SCT interaction, that is,

$$\text{NEG} = \beta_1\text{SAT} + \beta_2\text{ALT} + \beta_3\text{INV} + \beta_4\text{SCT} + \beta_7\text{INV}*\text{SCT} + \zeta, \qquad (6)$$

are shown in Table 4.3, and the EQS program code for this specification is shown in Fig. 4.2. The statistical significance of the INV*SCT interaction effect on NEG is shown in Table 4.4.

Discussion

The LISREL 8 coefficient estimates for Equation 5 shown in Table 4.1 were produced following the direct estimation procedure described previously. Each indicator for SAT, for instance, was centered by subtracting the indicator's average from its value in each case. The values for the single indicators

TABLE 4.2
Structural Model Results for the Single-Indicator Specification and LISREL 8

Structural Equation Analysis Estimates

Parameter	*Estimate*[a]	*Parameter*	*Variance Estimate*[a]	*Parameter*	*Estimate*[a]	*t Value*
λ_{s1}	0.559	ε_{s1}	0.166	ϕ_{SAT}	1.000	
λ_{s2}	0.625	ε_{s2}	0.130	ϕ_{ALT}	1.000	
λ_{s3}	0.699	ε_{s3}	0.110	ϕ_{INV}	1.000	
λ_{s4}	0.620	ε_{s4}	0.119	ϕ_{SCT}	1.000	
λ_{s5}	0.663	ε_{s5}	0.101	$\phi_{SAT:SCT}$	1.000	
λ_{a1}	0.843	ε_{a1}	0.272	$\phi_{ALT:SCT}$	1.000	
λ_{a2}	0.818	ε_{a2}	0.249	$\phi_{INV:SCT}$	1.000	
λ_{a3}	0.910	ε_{a3}	0.077	$\phi_{SCT:SCT}$	1.000	
λ_{a4}	0.713	ε_{a4}	0.244	$\phi_{SAT,ALT}$	−0.550*	
λ_{i1}	0.683	ε_{i1}	0.454	$\phi_{SAT,INV}$	0.347*	
λ_{i2}	0.752	ε_{i2}	0.120	$\phi_{SAT,SCT}$	0.257*	
λ_{i3}	0.766	ε_{i3}	0.089	$\phi_{ALT,INV}$	−0.278*	
λ_{i4}	0.750	ε_{i4}	0.124	$\phi_{ALT,SCT}$	−0.398*	
λ_{sc1}	0.884	ε_{sc1}	0.292	$\phi_{INV,SCT}$	0.532*	
λ_{sc2}	0.962	ε_{sc2}	0.209	ψ_{NEG}	0.319*	
λ_{sc3}	0.947	ε_{sc3}	0.175	$\gamma_{NEG,SAT}$	−0.241	−4.394
λ_{sc4}	0.979	ε_{sc4}	0.210	$\gamma_{NEG,ALT}$	0.132	2.398
$\lambda_{sat:sct}$	0.597	$\varepsilon_{sat:sct}$	0.046	$\gamma_{NEG,INV}$	−0.138	−2.601
$\lambda_{alt:sct}$	0.774	$\varepsilon_{alt:sct}$	0.087	$\gamma_{NEG,SCT}$	−0.018	−0.341
$\lambda_{inv:sct}$	0.696	$\varepsilon_{inv:sct}$	0.076	$\gamma_{NEG,SAT:SCT}$	0.120	2.762
$\lambda_{sct:sct}$	0.889	$\varepsilon_{sct:sct}$	0.203	$\gamma_{NEG,ALT:SCT}$	0.001	0.024
λ_{n1}	0.691	ε_{n1}	0.188	$\gamma_{NEG,INV:SCT}$	−0.118	−2.684
λ_{n2}	0.717	ε_{n2}	0.107	$\gamma_{NEG,SCT:SCT}$	−0.022	−0.482
λ_{n3}	0.865	ε_{n3}	0.063			
λ_{n4}	1.000	ε_{n4}	0.141			

Fit Indices

Chi-Square Statistic Value	681
Chi-Square Degrees of Freedom	269
p Value of Chi-Square Value	.000
GFI	.803
AGFI	.762
Comparative Fit Index	.910
Standardized RMS Residual	.088
RMSEA	.083
p Value for RMSEA <.05	.413E-06

TABLE 4.2
(Continued)

OLS Regression Estimates					
Dependent Variable	*Independent Variable*[b]	*b Coefficient*	*p Value*	*F Value and (p)*	R^2
NEG	SAT	−.290	.000	15.45(.000)	.367
	ALT	.137	.006		
	INV	−.170	.002		
	SWC	.004	.915		
	SAT*SCT	.158	.014		
	ALT*SCT	−.004	.941		
	INV*SCT	−.133	.021		
	SCT*SCT	−.022	.646		
	Constant	.048	.311		

[a]Maximum likelihood.
[b]The independent and dependent variables were averaged and centered.
*t value > 2.

TABLE 4.3
Structural Model Results for the Single-Indicator Specification and EQS

Structural Equation Analysis Estimates							
						t Value	
Parameter	*Estimate*[a]	*Parameter*	*Variance Estimate*[a]	*Parameter*	*Estimate*[a]	*ML*	*ML-Robust*
λ_{s1}	0.792	ε_{s1}	0.167	ϕ_{SAT}	0.518		
λ_{s2}	0.886	ε_{s2}	0.130	ϕ_{ALT}	0.849		
λ_{s3}	1.000	ε_{s3}	0.109	ϕ_{INV}	0.608		
λ_{s4}	0.879	ε_{s4}	0.119	ϕ_{SCT}	0.969		
λ_{s5}	0.940	ε_{s5}	0.102	$\phi_{INV:SCT}$	0.753		
λ_{a1}	0.926	ε_{a1}	0.271	$\phi_{SAT,ALT}$	−0.371*		
λ_{a2}	0.904	ε_{a2}	0.249	$\phi_{SAT,INV}$	0.197*		
λ_{a3}	1.000	ε_{a3}	0.077	$\phi_{SAT,SCT}$	0.188*		
λ_{a4}	0.783	ε_{a4}	0.244	$\phi_{SAT,INV:SCT}$	0.055		
λ_{i1}	0.893	ε_{i1}	0.452	$\phi_{ALT,INV}$	−0.203*		
λ_{i2}	0.986	ε_{i2}	0.120	$\phi_{ALT,SCT}$	−0.366*		
λ_{i3}	1.000	ε_{i3}	0.089	$\phi_{ALT,INV:SCT}$	−0.056		
λ_{i4}	0.978	ε_{i4}	0.124	$\phi_{INV,SCT}$	0.411*		
λ_{sc1}	0.902	ε_{sc1}	0.293	$\phi_{INV,INV:SCT}$	−0.171*		
λ_{sc2}	0.981	ε_{sc2}	0.211	$\phi_{SCT,INV:SCT}$	−0.020		
λ_{sc3}	0.967	ε_{sc3}	0.174	ψ_{NEG}	0.327*		
λ_{sc4}	1.000	ε_{sc4}	0.208	$\gamma_{NEG,SAT}$	−0.363	−4.671	−3.622
$\lambda_{inv:sct}$	0.928	$\varepsilon_{inv:sct}$	0.057	$\gamma_{NEG,ALT}$	0.163	2.697	−2.640
λ_{n1}	0.695	ε_{n1}	0.189	$\gamma_{NEG,INV}$	−0.172	−2.367	−2.092
λ_{n1}	0.695	ε_{n1}	0.189	$\gamma_{NEG,SCT}$	0.013	0.245	0.224
λ_{n3}	0.872	ε_{n3}	0.063	$\gamma_{NEG,INV:SCT}$	−0.111	−2.094	−2.403
λ_{n4}	1.000	ε_{n4}	0.141				

(Continued)

TABLE 4.3
(Continued)

Fit Indices	*ML*	*ML-Robust*
Chi-Square Statistic Value	321	276
Chi-Square Degrees of Freedom	196	
p Value of Chi-Square Value	.000	
GFI	.882	
AGFI	.848	
Comparative Fit Index	.971	.972
Standardized RMS Residual	.045	
RMSEA	.053	
p Value for RMSEA <.05	.268	
Squared Multiple Correlation for NEG = .372		

OLS Regression Estimates

Dependent Variable	*Independent Variable*[b]	*b Coefficient*	*p Value*	*F Value and (p)*	R^2
NEG	SAT	−.318	.000	22.68(.000)	.344
	ALT	.155	.001		
	INV	−.157	.005		
	SWC	.021	.623		
	INV*SCT	−.100	.031		
	Constant	.042	.284		

[a]Maximum likelihood.
[b]The independent and dependent variables were averaged and centered.
*t value > 2.

```
/TITLE=NEGLECT with INV*SCT only (Structural Model);
/SPECIFICATIONS
 VARIABLES = 53; ME = ml,robust;
 cases=222;MA = raw;
 FO='(f9.6,7f10.6/f9.6,7f10.6/f9.6,7f10.6/
      f9.0,6f10.0,f10.4/f9.4,7f10.4/
      f9.4,7f10.6/f9.6,4f10.6)';
 DATA='c:\eqswin\neg\neg1.dat';
/labels
 v2=sa2;v4=sa4;v5=sa5;v6=sa6;v7=sa7;v9=al2;v10=al3;v11=al4;v12=al5;
 v14=in1;v16=in3;v17=in4;v18=in5;v21=sc2;v22=sc3;v23=sc4;v24=sc5;
 v26=ne2;v29=ne5;v30=ne6;v31=ne7;v42=ssc;v45=asc;v47=isc;v48=scsc;
/EQUATIONS
 V2 = .9*f1            + e1;
 V4 = .9*f1            + e2;
 v5 = 1.0f1            + e3;
 v6 = .9*f1            + e4;
 v7 = .9*f1            + e5;
 v9 =   .9*f2          + e6;
 v10 =   .9*f2         + e7;
 v11 =   1.0f2         + e8;
```

FIG. 4.2. *(Continued)*

```
 v12 = .9*f2              + e9;
 v14 =    .9*f3           + e10;
 v16 =    .9*f3           + e11;
 v17 =   1.0f3            + e12;
 v18 =    .9*f3           + e13;
 v21 =      .9*f4         + e14;
 v22 =      .9*f4         + e15;
 v23 =      .9*f4         + e16;
 v24 =     1.0f4          + e17;
 v26 =        .9*f5       + e18;
 v29 =        .9*f5       + e19;
 v30 =        .9*f5       + e20;
 v31 =       1.0f5        + e21;
!v42 =   0.865307292f6 + e22;
!v45 =   0.870706444f7 + e23;
 v47 =   0.928009505f8 + e24;
!v48 =   0.928062490f9 + e25;
 f5 = -.29*f1 + .13*f2 + -.17*f3 + -.004*f4 + -.13*f8 + d1;
/VARIANCES
 F1 = .51985*;
 f2 = .8487*;
 f3 = .603*;
 f4 = .96864*;
!f5 =
!F6 = 0.539276064*;
!F7 = 0.956421678*;
 F8 = 0.753249384;
!F8 = 0.753249384*;
!F9 = 1.876526899*;
 e1 to e21 = .1*;
!e22 = 0.047264018;
!e23 = 0.033451716;
 e24 = 0.057338752;
!e25 = 0.205695870;
 d1 = .5*;
/COVARIANCE
 f1,f2 = -.37*;
 f1,f3 = .19*;
 f1,f4 = .18*;
 f2,f3 = -.20*;
 f2,f4 = -.36*;
 f3,f4 = .41*;
 f1,f8 = *;f2,f8 = *;f3,f8 = *;f4,f8 = *;
/print
 dig=5;
/END
```

FIG. 4.2. Single-Indicator EQS Specification Code

of the four interactions and the quadratic were added to each case. The single-indicator value for SAT*INV, for instance, was computed in each case using sat:neg = $[(sa_2 + sa_4 + sa_5 + sa_6 + sa_7)/5][(in_1 + in_3 + in_4 + in_5)/4]$. Next, the structural model was specified using PAR variables (Jöreskog & Sörbom, 1993), constraint equations (Jöreskog & Sörbom, 1993) for the loadings, errors, and variances for the interactions and the quadratic, and latent variable metrics assigned using unit variances for the exogenous variables (see Fig. 4.1 and Jöreskog & Sörbom, 1993); and the structural model was estimated using maximum likelihood. The use of PAR variables in this manner is sensitive to the sequence and location of the PAR and constraint (CO) statements in the program. In general PARs should not be used recursively (Jöreskog & Sörbom, 1993). In this application they should appear at the end of the program, just before the output (OU) card. In addition the PAR variables and the variables constrained in the CO statements (e.g., lx(18,5) in Fig. 4.1) should be defined in their natural numerical order (e.g., PAR(1), PAR(2), etc., not PAR(2), PAR(1); lx(18,5), lx(19,6), etc., not vice versa), and a PAR variable should be used in a CO statement as soon as possible after it is defined (see Fig. 4.1).

The EQS estimates for Equation 6 shown in Table 4.3 were produced using the procedure for two-step estimation described earlier. SAT, ALT, INV, SCT, and NEG were unidimensionalized. In this application unidimensionality was conceptualized as: The indicators of each construct have only one underlying construct each. Although there have been many proposals for developing unidimensional constructs (see Footnote 7), for this example unidimensionality was established as follows. Unidimensionality was operationalized as (a) for each construct the chi-square p value resulting from the single-construct measurement model (see Jöreskog, 1993) for each construct is .01 or greater; and (b) for the linear-terms-only measurement model (i.e., containing only the linear constructs) not rejecting H_0: the root mean square error of approximation < .05 (i.e., its p value > .05), and a comparative fit index of .95 or larger. The measurement model for SAT, for instance, was estimated using all of its items, then re-estimated until a target chi-square p value .01 or greater was attained by serially deleting items that did not appear to degrade content validity (not reported). This is obviously an area where judgment was required, and considerable care was taken to avoid sacrificing content validity for low chi-square. Because several itemizations for each construct were acceptable under criteria (a), judges were used to evaluate the content validity of each of these itemizations and select the final itemization for each construct. A linear-terms-only measurement model involving the final itemization for each construct suggested the constructs were unidimensional using the step (b) criteria (see Table 4.5).

Next, the indicators for SAT, ALT, INV, SCT, and NEG were centered, and the single indicator inv:sct = $[(in_1 + in_3 + in_4 + in_5)/4][(sc_2 + sc_3 + sc_4 +$

TABLE 4.4
INV-SCT Interaction Statistical Significance

SCT-NEG Assoc.				INV-NEG Assoc.			
INV Value[a]	*SCT Coefficient*[b]	*SE of SCT Coefficient*[c]	*t Value*	*SCT Value*[d]	*INV Coefficient*[e]	*SE of INV Coefficient*[f]	*t Value*
				1	0.07	0.11	0.66
2	0.21	0.11	1.83	2	–0.03	0.07	–0.42
3	0.10	0.07	1.38	3	–0.14	0.06	–2.08
3.80[g]	0.01	0.05	0.24	3.25[g]	–0.17	0.07	–2.36
4	–0.00	0.05	–0.15	4	–0.25	0.09	–2.70
5	–0.11	0.07	–1.51	5	–0.36	0.13	–2.68

[a]The values ranged from 2 (=low) to 5 in the study.
[b]The coefficient of SCT is given by (.013 – .111INV)SCT with INV centered.
[c]The standard error of the SCT coefficient is given by

$$\sqrt{Var(b_{SCT} + b_{INV*SCT}INV)} = \sqrt{Var(b_{SCT}) + INV^2 Var(b_{INV*SCT}) + 2INV Cov(b_{SCT}, b_{INV*SCT})}$$

[d]The values ranged from 1 (=low) to 5 in the study.
[e]The coefficient of INV is given by (–.172 – .111SCT)INV with SCT centered.
[f]The Standard Error of the INV coefficient is given by

$$\sqrt{Var(b_{INV} + b_{INV*SCT}SCT)} = \sqrt{Var(b_{INV}) + SCT^2 Var(b_{INV*SCT}) + 2SCT Cov(b_{INV}, b_{INV*SCT})}$$

[g]Mean value.

TABLE 4.5
Linear-Terms-Only Measurement Model Results

Parameter	*Estimate*[a]	*Parameter*	*Variance Estimate*[a]	*Parameter*	*Estimate*[a]
λ_{s1}	0.792	ε_{s1}	0.167	ϕ_{SAT}	0.517
λ_{s2}	0.886	ε_{s2}	0.130	ϕ_{ALT}	0.849
λ_{s3}	1.000	ε_{s3}	0.109	ϕ_{INV}	0.602
λ_{s4}	0.879	ε_{s4}	0.119	ϕ_{SCT}	0.968
λ_{s5}	0.940	ε_{s5}	0.102	ϕ_{NEG}	0.520
λ_{a1}	0.926	ε_{a1}	0.271	$\phi_{SAT,ALT}$	–0.371*
λ_{a2}	0.904	ε_{a2}	0.249	$\phi_{SAT,INV}$	0.199*
λ_{a3}	1.000	ε_{a3}	0.077	$\phi_{SAT,SCT}$	0.188*
λ_{a4}	0.783	ε_{a4}	0.244	$\phi_{SAT,NEG}$	–0.286*
λ_{i1}	0.891	ε_{i1}	0.454	$\phi_{ALT,INV}$	–0.205*
λ_{i2}	0.986	ε_{i2}	0.120	$\phi_{ALT,SCT}$	–0.366*
λ_{i3}	1.000	ε_{i3}	0.089	$\phi_{ALT,NEG}$	0.309*
λ_{i4}	0.978	ε_{i4}	0.124	$\phi_{INV,SCT}$	0.411*
λ_{sc1}	0.902	ε_{sc1}	0.293	$\phi_{INV,NEG}$	–0.118*
λ_{sc2}	0.982	ε_{sc2}	0.210	$\phi_{SCT,NEG}$	–0.184*
λ_{sc3}	0.967	ε_{sc3}	0.175		
λ_{sc4}	1.000	ε_{sc4}	0.209		
λ_{n1}	0.694	ε_{n1}	0.190		
λ_{n1}	0.730	ε_{n1}	0.107		
λ_{n3}	0.872	ε_{n3}	0.063		
λ_{n4}	1.000	ε_{n4}	0.141		

(Continued)

TABLE 4.5
(Continued)

Fit Indices	
Chi-Square Statistic Value	305
Chi-Square Degrees of Freedom	179
p Value of Chi-Square Value	.000
GFI	.883
AGFI	.850
Comparative Fit Index	.970
Standardized RMS Residual	.045
RMSEA	.056
p Value for RMSEA <.05	.153

[a]Maximum likelihood.
*t value > 2.

sc_5)/4] was added to each case. Using the linear-terms-only measurement model parameter estimates (see Table 4.5) the loading and error variance of inv:sct plus the variance of INV:SCT were calculated using Equations 3 and 1a. The structural model then was specified with the loading of inv:sct and its error variance fixed at the values calculated with Equation 3, and the variance of INV*SCT fixed at the Equation 1a value. Specifically, the loading of inv:sct was fixed at $\Gamma_{INV}\Gamma_{SCT}$ = .9280, the error variance of inv:sct was fixed at $\Gamma_X{}^2\text{Var}(X)\theta_Z + \Gamma_Z{}^2\text{Var}(Z)\theta_X + \theta_X\theta_Z$ = .0573, and the variance of INV*SCT was fixed at $\text{Var(INV)Var(SCT)} + \text{Cov(INV,SCT)}^2$ = .7532 (see Fig. 4.2). Finally, the structural model was estimated using maximum likelihood estimation in EQS.

Indicator Non-Normality. Whereas the measurement parameter, structural disturbance, and coefficient estimates (e.g., λs, θs, ϕs, ψs, γs, and βs) produced by maximum likelihood are robust to departures from normality (T. W. Anderson & Amemiya, 1985, 1986; Bollen, 1989; Boomsma, 1983; Browne, 1987; Harlow, 1985; Jöreskog & Sörbom, 1989; Sharma, Durvasula, & Dillon, 1989; Tanaka, 1984), the model fit and significance statistics (i.e., chi-square and standard errors) may not be (Bentler, 1989; Bollen, 1989; Jöreskog & Sörbom, 1989) (see Jaccard & Wan, 1995, for evidence to the contrary). Because product and single indicators are per se non-normal, and model fit and significance statistics from the maximum likelihood-robust estimator (Satorra & Bentler, 1988) appear to be less sensitive to departures from normality (see Hu, Bentler, & Kano, 1992), maximum likelihood-robust estimates were added to the Table 4.3 results.[10] However, comparing the

[10]Asymptotic distribution free estimates (WLS and DWLS methods in LISREL and the AGLS method in EQS) appear to be inappropriate for small samples (i.e., fewer than 400 cases) (Aiken & West, 1991; Hu et al., 1992; Jaccard & Wan, 1995) (see also Jöreskog & Yang, 1996).

standard error and chi-square estimates for maximum likelihood and maximum likelihood-robust, they were generally similar. The limited evidence to date (Jaccard & Wan, 1995; Ping, 1995) suggests that for most practical purposes (Jöreskog & Yang, 1996) model fit and significance statistics from maximum likelihood estimation may be sufficiently robust to the addition of a few product indicators that involve linear indicators that are normal. However, the robustness of these statistics from this estimator to the addition of many product indicators (i.e., over four) or product indicators comprised of non-normal linear indicators typical of survey data is unknown.

Reestimation. Comparing the Table 4.3 and 4.5 measurement parameter estimates, they are similar. Had they been dissimilar (i.e., different in the second decimal place) the re-estimation process could have been used. Using the re-estimation technique, the requirement for strict unidimensionality in the linear latent variables can be relaxed somewhat, although the practical limits of how different the measurement parameters can be between the linear-terms-only measurement model and the structural model to produce stable measurement and structural coefficient estimates is unknown.

*Interpreting INV*SCT.* Table 4.4 provides information regarding the contingent nature of the SCT relationship with NEG (and the INV relationship with NEG). The statistical significance of the SCT coefficient (i.e., .013 – .111INV) varies with INV: The size of the coefficient and its standard error depend on the level of INV. In addition, the standard error of the coefficient of SCT involves the variance and covariance of the INV and SCT*INV coefficients. Table 4.4 also demonstrates the effect of an interaction. When INV is at its study average, the SCT coefficient was small and nonsignificant. For smaller values of INV it was positive and approached significance. As this example suggests, had the interaction been significant at both ends of the range of INV, offering plausible explanations for disordinal interactions can be challenging (see Aiken & West, 1991).[11]

Intercepts. In realistic social science research situations with centered indicators for all the latent variables, the omission of an intercept term in a structural equation with an interaction or quadratic biases the resulting coefficients slightly (see Jöreskog & Yang, 1996).[12] In these situations this

---

[11]An *XZ* interaction can be ordinal or disordinal (Lubin, 1961). For an ordinal interaction the *Z-Y* association becomes weaker over the range of the interacting variable *X*. A disordinal interaction, however, is characterized by the *Z-Y* association changing signs over the range of the interacting variable *X*.

[12]As Jöreskog and Yang (1996) pointed out, a structural equation containing an interaction should in general be specified with an intercept term to avoid interpretational difficulties. Neglecting to do so biases structural coefficients and their standard errors, and produces an interaction coefficient that represents a centered interaction, which may be difficult to interpret. However, the specification of

bias is typically in the third or fourth decimal place. For instance, the Table 4.3 results with an intercept were

$$\text{NEG} = .043 - .363\text{SAT} + .163\text{ALT} - .173\text{INV} - .013\text{SCT} - .116\text{INV*SCT}.^{13}$$
$$t = \quad 1.06 \quad -4.68 \quad 2.69 \quad -2.41 \quad .25 \quad -2.26$$

As a result, unless mean structures are of interest, it is usually unnecessary to estimate intercepts with latent variable interactions and quadratics, and the Table 4.4 results used the Table 4.3 coefficient and standard error estimates, which assumed a zero intercept.

However, for any structural coefficient in a neighborhood of twice its maximum likelihood standard error (e.g., between $|\mathbf{t}| = 1.85$ and 2.15), the intercept-influenced coefficient should be estimated to avoid Type I and II errors. Because modeling intercepts can produce model identification, convergence, and improper solution problems (see Bentler, 1989), and adding interactions and quadratics frequently aggravate these problems, an intercept model containing interactions and quadratics could be estimated in two steps. First the interaction and quadratic model is estimated as described earlier (i.e., without an intercept). Then the model is re-estimated with the intercept(s) specified, using starting values from the no-intercept model and ordinary least squares (OLS) regression estimates of the intercept(s). If estimation problems are encountered in the second step, one or more structural coefficients for linear latent variables not involved in the interactions and quadratics could be fixed at their no-intercept values to obtain starting values for the other structural coefficients and the intercept. For larger models, however, it may be impossible to estimate an intercept model with interactions, quadratics, and centered linear latent variables. For unknown reasons the Equation 5 model with an intercept would not converge using any of the techniques discussed previously. OLS regression results for Equation 5 with an intercept and those for regression-through-the-origin (which is equivalent to the Equation 5 model) were

$$\text{NEG} = .048 - .290\text{SAT} + .137\text{ALT} - .170\text{INV} - .004\text{SCT} + .158\text{SAT*SCT} - .004\text{ALT*SCT} - .133\text{INV*SCT} - .022\text{SCT*SCT}$$

and

an intercept along with interactions and quadratics can create estimation difficulties, and limit the accessibility of estimators that are less sensitive to departures from normality. Because no-intercept bias is substantially reduced in realistic social science research situations by centering the indicators of the independent and dependent latent variables, this presentation centers all indicators and reports estimation results that typically omit the specification of intercepts.

[13]These estimates were produced by adding METHOD=MOMENT to the /SPECIFICATIONS section of Fig. 4.2, and adding a *V999 variable to the equation for F5.

$$NEG = -.299SAT + .137ALT - .168INV - .001SCT + .160SAT*SCT - .002ALT*SCT - .132INV*SCT + .005SCT*SCT.$$

Comparing these coefficient estimates suggests that any intercept bias in the Fig. 4.1 results may be small.

Alternatives. In the first example, two-step estimation with single indicators could have been used instead of direct estimation. The estimation procedure would have been the same as in the second example except that the computed starting values for the loadings and error variances of the interactions and the quadratic (plus their variances if the diagonal ϕs had not equaled 1) would have been fixed. Both direct and two-step estimation using single indicators are useful for probing for the existence of interactions or quadratics because they are relatively easy to specify, and both produce sample covariance matrices with fewer elements than the Kenny and Judd (1984) approach. The single-indicator direct approach works well in LISREL 8 for smaller models, but because the PARs are used recursively, larger models can misbehave. For instance, the standard errors and statistical significance for several interaction loadings and errors did not print in the first example. Eliminating the single-indicator PARs by replacing them with their expanded equivalents does not usually execute in larger models such as the first example.

Similarly, the second example could have used product indicators with either direct or indirect estimation. Specification effort would have been higher than for a single-indicator model if the full set of product indicators were used for each nonlinear latent variable (to avoid concerns about their content validity—discussed later). For the calculations involved in fixed and/or starting values for product or single indicators, an Excel or Lotus spreadsheet is useful. An Excel spreadsheet was used for starting values in both examples, and for the fixed single-indicator loading and error variance, plus the fixed variance of INV*SCT, in the second example (see Table 4.6). The linear-terms-only measurement model loadings, error variances, and variances and covariances for the linear latent variables involved in the interactions and quadratic (e.g., SAT, ALT, INV, and SCT; see Table 4.5) were keyed into the spreadsheet, and the product indicators' loadings and error variances, plus the interaction and quadratic variances, were calculated using Equations 1 through 4.

CONVERGENCE AND PROPER ESTIMATES

These techniques can produce their share of convergence and improper solution headaches. As with any structural equation model solution, the output should be examined for negative squared multiple correlations, linearly dependent parameter estimates, parameter estimates constrained at zero, and

TABLE 4.6
Spreadsheet for the Single-Indicator Loadings and Errors

A	*B*	*C*	*D*	*E*	*F*	*G*	
1 EXCEL spreadsheet to calculate interaction and quadratic fixed and starting values							
2 using linear terms only measurement model values (see Table 4)							
3							
4 Unstandardized Linear-Terms-Only Measurement Model Values:							
5 Lambda						Sum	
6 SAT	0.7923	0.88286	1	0.87807	0.93786	0.898218	(=SUM(B6:F6)/5)
7 ALT	0.92619	0.90489	1	0.78421		0.9038225	(=SUM(B7:E7)/4)
8 INV	0.8907	0.9848	1	0.97772		0.963305	(=SUM(B8:E8)/4)
9 SCT	0.90298	0.98275	0.96771	1		0.96336	(=SUM(B9:E9)/4)
10 Theta:						Sum	
11 SAT	0.16612	0.13209	0.10738	0.11977	0.10302	0.025135	(=SUM(B11:F11)/5^2)
12 ALT	0.27292	0.24845	0.0784	0.24354		0.052707	(=SUM(B12:E12)/4^2)
13 INV	0.45457	0.12132	0.08869	0.12417		0.049297	(=SUM(B13:E13)/4^2)
14 SCT	0.29309	0.21006	0.17531	0.2094		0.055491	(=SUM(B14:E14)/4^2)
15 Phi:							
16	SAT	ALT	INV	SCT			
17 SAT	0.51985						
18 ALT	–0.37181	0.8487					
19 INV	0.19978	–0.20527	0.603				
20 SCT	0.18902	–0.36652	0.41129	0.96864			
Calculated Interaction and Quadratic Values:							

(Continued)

TABLE 4.6
(Continued)

	Lambda	
sat:sct	0.865307292	(=+G6*G9)
alt:sct	0.870706444	(=+G7*G9)
inv:sct	0.928009505	(=+G8*G9)
sct:sct	0.92806249	(=+G9^2)
	Theta	
sat:sct	0.233968111	(=+G6^2*B17*G14+G9^2*E20*G11+G11*G14)
alt:sct	0.16890411	(=+G7^2*B18*G14+G9^2*E20*G12+G12*G14)
inv:sct	0.262181552	(=+G8^2*B19*G14+G9^2*E20*G13+G13*G14)
sct:sct	0.896686172	(=4*G9^2*E20*G14+2*G14^2)
	Phi	
sat:sct	0.539276064	(=+B17*E20+B20^2)
alt:sct	0.956421678	(=+C18*E20+C20^2)
inv:sct	0.753249384	(=+D19*E20+D20^2)
sct:sct	1.876526899	(=2*E20^2)

so on. When convergence or improper solution difficulties are encountered using the techniques discussed earlier, the first step should be to verify that the indicators of the linear latent variables were centered.

User-specified starting values for the latent variable variances and covariances, the structural coefficients (i.e., the γs and βs), and the variances of the structural disturbances (i.e., Var(ζs) are frequently required. Error-attenuated variance and covariance estimates from SAS or SPSS, and OLS regression coefficient estimates are frequently sufficient to solve convergence and improper solution problems. The estimated disturbance term variance for Y, for instance, can be calculated using $\text{Var}(\zeta) = \text{Var}(Y)(1 - R^2)$, where Var($Y$) is the error-attenuated variance of Y (e.g., from SAS or SPSS) and R^2 is the OLS regression estimate of the explained variance for Y regressed on the summated linear (e.g., $(x_1 + x_2)/2$), interaction, and quadratic variables (e.g., $[(x_1 + x_2)/2][(z_1 + z_2 + z_3)/3]$) involved in the structural equation model.

Occasionally, disattenuated variance and covariance estimates are required. The adjusted variance and covariance estimates for the independent variables can be calculated using attenuated variances and covariances (i.e., SAS or SPSS estimates), linear-terms-only measurement model estimates, and the calculations shown in the Appendix. As an alternative a measurement model involving all the latent variables in the structural model of interest (e.g., X, Z, Y, XZ, and XX) should also provide useful disattenuated variance and covariance estimates.

If problems persist, constraining the variance of the structural disturbance terms (i.e., ζ in Equations 5 or 6) to more than 10% of the variance of their respective endogenous variables may be effective (few models in the social sciences explain 90% of the variance of an endogenous variable). This would be accomplished in the first example by constraining PSI(1), and in the second example by constraining d1.

In addition, scaling the exogenous variables by setting their variance to 1 (see Jöreskog & Sörbom, 1993) is sometimes useful. This approach was used in the first example: The loadings of SAT, ALT, INV, and SCT were all freed, and their variances plus the variances of the interactions and the quadratic were fixed at 1 (see Fig. 4.1).

The Equation 1a and 2a constraints on the variances of the interactions and quadratics could also be relaxed in problem situations. This would be accomplished in the first example by deleting the constraint equations for the variances of the interactions and quadratic. In the second example, the fixed variances for the interaction would be freed by adding a *. The resulting interaction or quadratic coefficient estimates are typically attenuated and closer to their OLS regression estimates.

In addition, fixing some of the covariances among the exogenous variables at zero is occasionally necessary. In particular, zeroing the covariances between the linear latent variables (e.g., X and Z) and the interactions/quadratics (e.g., XZ), and/or zeroing the covariances among the interactions/quadratics (e.g., between XZ and XX) may be required. This was done in the first example to compensate for the multicollinearity among the interactions and the quadratic.

If problems continue to persist, the model may not be identified. Otherwise identified models can become nonidentified with the specification of correlated errors or nonrecursive relationships (see Berry, 1984). More often, with interactions and quadratics, otherwise identified models can be empirically underidentified, or weakly identified (see Hayduk, 1987). Berry (1984), Bollen (1989), and Hayduk (1987) provided accessible discussions of the sources of lack of identification and identification checking.

NEEDED RESEARCH

Although the aforementioned techniques provide considerable improvement over regression coefficient estimates for tests of theory involving unobserved interaction and quadratic variables, additional work is needed on the specification, estimation, and interpretation of these variables using structural equation analysis. The following is an incomplete enumeration of areas where additional research on these matters might be useful, in no particular order of importance.

Specification

The number of product indicators in a model with interactions and quadratics involving overidentified constituent latent variables can become large. Equation 5 for instance would have required 62 product indicators. As mentioned earlier, specifying many product indicators can add to execution times, convergence, and improper solution problems. However, a reduced number of product indicators may adequately specify a latent variable interaction or quadratic. Jaccard and Wan (1995), for instance, used a subset of four product indicators.

It could be argued, however, that concern for the content validity of a latent variable interaction or quadratic requires the use of all its product indicators. All of the indicators of the interaction and quadratic latent variables were used in the examples presented earlier because it was not clear which product indicators could have been safely dropped without impairing the content validity of interactions and the quadratic. As a result, it would be useful to know the conditions under which product indicators could safely be dropped without impairing the content validity of the resulting interaction or quadratic latent variable, or to have guidelines for this endeavor.

In an investigation of spurious interactions in regression, Lubinski and Humphreys (1990) noted that interactions and quadratics are correlated. As the correlation between X and Z, for instance, approaches 1, the correlation between XX (or ZZ) and XZ also approaches 1. As a result, they argued a population quadratic can be mistaken for an interaction in regression. Consequently, they proposed that quadratic combinations of the linear latent variables comprising a latent variable interaction should be entered in the regression model along with the interaction of interest.

It is plausible that these results may extend to structural equation analysis. In the first example, the quadratic was included because it was a second-order latent variable involving switching cost. Lubinski and Humphreys' (1990) results suggest that it should have been included also because a significant INV*SCT interaction could be induced by significant SCT*SCT or INV*INV quadratics in the population equation. Consequently, it would be helpful to know if latent variable interactions and quadratics can be mistaken for each other in structural equation analysis, and if so, what remediation steps would be appropriate.

Kenny and Judd (1984) proposed that, under their normality assumptions, the variances of latent variable interactions and quadratics should be constrained to their respective Equation 1a and 2a forms, which are based on Kendall and Stuart's (1958) results. Kendall and Stuart also showed under these conditions that interactions and quadratics are associated with each other and linear latent variables in a predictable manner (e.g., $\text{Cov}(XZ,XX) = 2\text{Var}(x)\text{Cov}(X,Z)$, $\text{Cov}(XZ,X) = 0$, etc.). However these constraints were not specified in Kenny and Judd, and they have not been specified subsequently (see Jaccard & Wan, 1995; Jöreskog & Yang, 1996; Ping, 1995, 1996a). For instance, in the first example, the covariances of the interactions with each other and the quadratic were not constrained, nor were their covariances with the linear latent variables constrained. Although Jaccard and Wan's results suggest that the omission of Cov(XZ,*) constraints, where * is a linear latent variable, may not materially affect interaction coefficients, their study may not have been designed to investigate this matter. As a result, it would be interesting to know what effect, if any, the omission of Cov(XZ,*) constraints, and Cov(XZ,**) constraints (when an interaction and a quadratic, and/or multiple interactions and/or quadratics are jointly specified) have on the resulting interaction and quadratic coefficients.

It is not obvious how a nonrecursive latent variable interaction or quadratic should be estimated. A nonrecursive interaction or quadratic specification may be appropriate when a hypothesized feedback relationship is not linear in one or both directions. This situation is plausible in the examples presented previously. It could be argued that investment in a relationship should reduce relationship neglect, and neglect should also reduce investment. Assuming the neglect-investment relationship would still be moderated by switching cost (see Table 4.4) in a nonrecursive specification of this relationship, and recalling the requirements for identification and instrumental variables that are not correlated with their indirect endogenous variables in these specifications (see Berry, 1984), it would be useful to know how to specify this relationship in a structural equation model using a minimum of additional nonrecursive paths and instrumental variables.

It would also be helpful to have some guidance regarding the estimation of mixed formative and reflexive models involving interactions and quadratics. Although adequate techniques such as partial least squares (Lohmöller, 1981) exist for estimating these models (see Fornell & Bookstein, 1982), substantive researchers frequently use OLS regression when their structural model contains a mixture of formative and reflexive variables (see, e.g., Heide & John, 1990; however, see Bristor, 1993). As a result, at least some of the structural coefficient estimates are biased and inefficient (Bohrnstedt & Carter, 1971). Hence, it would be useful to know how to specify formative interactions and quadratics, and mixed formative-reflexive interactions.

Estimation

A hierarchical procedure for sequentially adding interactions and quadratics to a model such as that used in ANOVA or hierarchical regression analysis would also be useful to avoid interaction or quadratic latent variables that are significant but explain little additional variance and are therefore of little substantive value. The second example is a case in point. An Equation 6 model that excluded the INV*SCT interaction explained 35.6% of the variance of NEG (not reported). This is slightly more than 1 percentage point less than the multiple squared correlation for NEG shown in Table 4.3 with INV*SCT specified, which would be nonsignificant in a hierarchical regression analysis.

Research design could affect the detection of a latent variable interaction or quadratic. In an exploration of the difficulties of detecting interactions using survey data and regression, McClelland and Judd (1993) showed that because field studies are similar to an experiment with unequal cell sizes, field studies are generally less efficient than experiments in detecting interactions (see also Stone-Romero, Alliger, & Aguinis, 1994). They concluded an optimal experimental design for detecting interactions exhibits an independent variable distribution that is polar (i.e., has many cases containing extreme independent variable values) and balanced (i.e., has equal cell sizes). The most efficient of these distributions McClelland and Judd characterized as a "four-cornered" data distribution (which has, for two independent variables with more than two levels, a three-dimensional frequency distribution that looks like the four legs on an upside-down kitchen table), and an "*X*-model" (which has, for two independent variables with more than two levels, a three-dimensional frequency distribution that resembles a bas-relief *X* anchored on the four polar cells).

Because field studies in the social sciences typically produce censored mound-shaped distributions for independent variables instead of uniform (balanced), four-cornered, or *X* distributions, they are not usually as efficient as experiments in detecting interactions. Comparing an experiment with two independent variables and a four-cornered data distribution to the equivalent mound-shaped field study distribution, McClelland and Judd (1993) argued that the field study would produce a data distribution that is 90% to 94% less efficient in detecting interactions as the four-cornered distribution.

Because it is plausible that their results extend to structural equation analysis, it would be helpful to have guidelines for nonexperimental research designs that would produce a high likelihood of detecting an hypothesized population interaction or quadratic. For instance, McClelland and Judd (1993) suggested oversampling the extremes or poles of the scales in such studies. Based on their results, a stratified sample that produces a uniform distribution for two independent variables increases the efficiency of detecting an interaction between these two variables using regression by a factor of between 2.5 and 4. A field study that samples only the poles, such as Dwyer and Oh's (1987) study of output sector munificence effects in marketing channels, improves the efficiency of interaction detection using regression by 1,250% to 1,666% (although they did not test for interactions).

It would also be useful to have suggestions regarding analytical technique refinements that would increase the likelihood of detecting a hypothesized population interaction or quadratic using a field survey design. Possibilities might include a case-weighting approach that emphasizes the polar cases in a set of responses, so that a more nearly uniform or polar distribution would be produced.

There may be other data-related factors that affect the detection of interactions and quadratics in structural equation analysis. For instance, correlated (systematic) error between the independent and dependent variables attenuates the observed coefficient sizes of interactions in regression (Evans, 1985). In the examples presented earlier, it was assumed that the effects of these errors were adequately modeled with uncorrelated indicator error terms. Although the techniques discussed earlier can be extended to models involving correlated errors for the linear latent variables (see Ping, 1995, 1996a), it would be helpful to have guidance for any implications this has for correlations among the error terms for product indicators.

Interpretation

As mentioned earlier, product and single indicators are not normal even if their constituent variables are normal. However the evidence to date (Jaccard & Wan, 1995; Kenny & Judd, 1984; Ping, 1995, 1996a), including the results presented in the second example, suggest the addition of a few of these indicators does not materially bias standard errors or the chi-square statistic of the resulting structural model using maximum likelihood. As a result, it would be useful to know the limits of these results. This may be important to substantive researchers because, when compared with maximum likelihood, estimators that are less dependent on distributional assumptions are practically unknown to them. Although studies involving Monte Carlo analyses with realistic research situations would be valuable (see, e.g., Jaccard & Wan, 1995), it would be interesting to see results derived from bootstrap techniques (however, see Bollen & Stine, 1993).

The techniques discussed in this chapter and elsewhere (e.g., Aiken & West, 1991; Jaccard et al., 1990) have been applied exclusively to exogenous variable interactions and quadratics. Although these techniques may extend with little modification to endogenous interactions and quadratics, it is not clear what a dependent interaction represents. Such a situation could arise in the examples presented previously. It is plausible that satisfaction is an antecedent of both investment and switching cost: As overall relationship satisfaction increases, investments in the relationship should increase, and the perception of the difficulty (cost) of switching relationships should also increase. However, if INV and SCT still combine (interact) in their relationship with NEG with a SAT antecedent, it is not clear how to conceptualize or interpret the SAT-to-NEG via INV, SCT and INV*SCT relationship.

Although model fit assessment is a controversial area (see Bollen & Long, 1993), guidelines for model fit when interactions and quadratics are specified would be useful. It has been my experience with field survey data and simulated data closely mimicking field survey data that as the number of specified product indicators for a significant interaction or quadratic increases, some model fit indices suggest model-data fit is improved, whereas others suggest the opposite. The addition of a single indicator for a significant interaction or quadratic also appears to produce conflicting model-to-data fit results, even when compared with a misspecified model that excludes a population interaction or quadratic.

It may also be helpful to revisit the reporting and interpretation of interactions and quadratics from a theory-testing perspective. Though Aiken and West (1991) provide an accessible treatment of this topic for interactions, there is no equivalent treatment of quadratics. In addition, the fact that the SCT effect on NEG, for instance, depends on the range and mean of INV, and each of these statistics has a confidence interval, suggests that the Table 4.4 presentation may be simplistic for conclusions regarding a theory test. It could be argued that in other contexts the significant disordinal interaction may be ordinal, and the SCT effect could be positive or negative over the range of INV. To explain, in a different study the range of the interacting variable INV could be different, the sample could produce a different mean for INV, and the observed interaction could be ordinal and positive, ordinal and negative, or disordinal and both. Hence, the most that might be concluded from the second example is there is an interaction between INV and SCT in their association with NEG, and the SCT-NEG effect could be positive, negative, or both over the range of INV.

Standardized structural coefficients are used in some disciplines (e.g., marketing) to compare the relative impact of latent variables with significant coefficients. However, standardized regression coefficients for interactions and quadratics are not invariant to centering (see Aiken & West, 1991), and comparisons among standardized coefficients in models involving interactions and quadratics that also involve centered linear variables may be misleading. Friedrich (1982) proposed using Z scores to produce standardized coefficients in regression involving interactions (see Aiken & West, 1991), and it would be useful to have equivalent results for interaction and quadratic latent variables.

Finally, it may be helpful to revisit the interpretation of an interaction when the coefficient for the linear variable is nonsignificant. The current practice is to interpret the first derivative of the combination of the nonsignificant linear variable coefficient and the significant interaction variable as shown in Table 4.4 (Aiken & West, 1991; Jaccard et al., 1990). It could be argued, however, that because the linear variable coefficient is nonsignificant, only the contribution of the interaction variable should be interpreted. For instance, in Table 4.4 the first derivative (.013 – .111INV) was interpreted. However, because the constant term in this first derivative is nonsignificant, should only the expression –.111INV*SCT be interpreted? In the Table 4.4 case, the SCT coefficient would have been significant at lower and higher values of INV (not reported) and the conclusions would not be the same as that suggested by the Table 4.4 results.

There may be other concerns as well. For instance, noninterval data analyzed as interval data produces biased estimates (Jöreskog & Sörbom,

1989) and the situation may or may not be aggravated by the specification of interactions and quadratics comprised of ordinal latent variables. In addition, the limits of departures from unidimensionality for the re-estimation technique to work in two-step estimation are not known. In summary, it is likely there are many needed useful additions to what is already known in this emerging area.

REFERENCES

Aaker, D. A., & Bagozzi, R. P. (1979). Unobservable variables in structural equation models with an application in industrial selling. *Journal of Marketing Research, 16,* 147–158.

Aiken, L. S., & West, S. G. (1991). *Multiple regression: Testing and interpreting interactions.* Newbury Park, CA: Sage.

Anderson, J. C., & Gerbing, D. W. (1982). Some methods for respecifying measurement models to obtain unidimensional construct measurement. *Journal of Marketing Research, 19,* 453–460.

Anderson, J. C., & Gerbing, D. W. (1988). Structural equation modeling in practice: A review and recommended two-step approach. *Psychological Bulletin, 103,* 411–423.

Anderson, T. W., & Amemiya, Y. (1985). The asymptotic normal distribution of estimators in factor analysis under general conditions (Tech. Rep. No. 12). Stanford, CA: Stanford University, Econometric Workshop.

Anderson, T. W., & Amemiya, Y. (1986). Asymptotic distribution in factor analysis and linear structural relations (Tech. Rep. No. 18). Stanford, CA: Stanford University, Econometric Workshop.

Bagozzi, R. P. (1992). State versus action orientation and the theory of reasoned action: An application to coupon usage. *Journal of Consumer Research, 18,* 505–518.

Bentler, P. M. (1989). *EQS structural equations program manual.* Los Angeles: BMDP Statistical Software.

Berry, W. D. (1984). *Nonrecursive causal models.* Beverly Hills, CA: Sage.

Blalock, H. M., Jr. (1965). Theory building and the concept of interaction. *American Sociological Review, 30,* 374–381.

Bohrnstedt, G. W., & Carter, T. M. (1971). Robustness in regression analysis. In H. L. Costner (Ed.), *Sociological methodology* (pp. 118–146). San Francisco: Jossey-Bass.

Bollen, K. A. (1989). *Structural equations with latent variables.* New York: Wiley.

Bollen, K. A., & Long, J. S. (1993). *Testing structural equation models.* Newbury Park, CA: Sage.

Bollen, K. A., & Stine, R. A. (1993). Bootstrapping goodness of fit measures in structural equation models. In K. A. Bollen & J. S. Long (Eds.), *Testing structural equation models* (pp. 111–135). Newbury Park, CA: Sage.

Boomsma, A. (1983). *On the robustness of LISREL (maximum likelihood estimation) against small sample size and nonnormality.* Unpublished doctoral dissertation, University of Groningen, Groningen, The Netherlands.

Bristor, J. M. (1993). Influence strategies in organizational buying: The importance of connections to the right people in the right places. *Journal of Business-to-Business Marketing, 1*(1), 63–98.

Browne, M. W. (1987). Robustness of statistical inference in factor analysis and related models. *Biometrika, 74,* 375–384.

Burt, R. S. (1973). Confirmatory factor-analysis structures and the theory construction process. *Sociological Methods and Research, 2,* 131–187.

Cohen, J. (1968). Multiple regression as a general data-analytic system. *Psychological Bulletin, 70,* 426–443.

Cohen, J., & Cohen, P. (1975). *Applied multiple regression/correlation analyses for the behavioral sciences.* Hillsdale, NJ: Lawrence Erlbaum Associates.

Cohen, J., & Cohen, P. (1983). *Applied multiple regression/correlation analyses for the behavioral sciences.* Hillsdale, NJ: Lawrence Erlbaum Associates.

Cronbach, L. J. (1987). Statistical tests for moderator variables: Flaws in analyses recently proposed. *Psychological Bulletin, 103,* 414–417.

Dwyer, F. R., & Oh, S. (1987). Output sector munificence effects on the internal political economy of marketing channels. *Journal of Marketing Research, 24,* 347–358.

Evans, M. T. (1985). A Monte Carlo study of the effects of correlated methods variance in moderated multiple regression analysis. *Organizational Behavior and Human Decision Processes, 36,* 305–323.

Fornell, C., & Bookstein, F. L. (1982). Two structural equation models: LISREL and PLS applied to exit-voice theory. *Journal of Marketing Research, 19,* 440–452.

Friedrich, R. J. (1982). In defense of multiplicative terms in multiple regression equations. *American Journal of Political Science, 26,* 797–833.

Gerbing, D. W., & Anderson, J. C. (1985). The effects of sampling error and model characteristics on parameter estimation for maximum likelihood confirmatory factor analysis. *Multivariate Behavioral Research, 20,* 255–271.

Gerbing, D. W., & Anderson, J. C. (1988). An updated paradigm for scale development incorporating unidimensionality and its assessment. *Journal of Marketing Research, 25,* 186–192.

Hanushek, E. A., & Jackson, J. E. (1977). *Statistical methods for social scientists.* New York: Academic Press.

Harlow, L. L. (1985). *Behavior of some elliptical theory estimators with nonnormal data in a covariance structures framework: A Monte Carlo study.* Unpublished doctoral dissertation, University of California, Los Angeles.

Hattie, J. (1985). Methodology review: Assessing unidimensionality of tests and items. *Applied Psychological Measurement 9,* 139–164.

Hayduk, L. A. (1987). *Structural equation modeling with LISREL: Essential and advances.* Baltimore: Johns Hopkins University Press.

Heide, J. B., & John, G. (1990). Alliances in industrial purchasing: The determinants of joint action in buyer–seller relationships. *Journal of Marketing Research, 27,* 24–36.

Howard, J. A. (1989). *Consumer behavior in marketing strategy.* Englewood Cliffs, NJ: Prentice-Hall.

Hu, L., Bentler, P. M., & Kano, Y. (1992). Can test statistics in covariance structure analysis be trusted? *Psychological Bulletin, 112,* 351–362.

Jaccard, J., Turrisi, R., & Wan, C. K. (1990). *Interaction effects in multiple regression.* Newbury Park, CA: Sage.

Jaccard, J., & Wan, C. K. (1995). Measurement error in the analysis of interaction effects between continuous predictors using multiple regression: Multiple indicator and structural equation approaches. *Psychological Bulletin, 117*(2), 348–357.

Jöreskog, K. G. (1970). A general method for analysis of covariance structures. *Biometrika, 57,* 239–251.

Jöreskog, K. G. (1971a). Simultaneous factor analysis in several populations. *Psychometrika, 57,* 409–426.

Jöreskog, K. G. (1971b). Statistical analysis of sets of congeneric tests. *Psychometrika, 36,* 109–133.

Jöreskog, K. G. (1993). Testing structural equation models. In K. A. Bollen & J. S. Long (Eds.), *Testing structural equation models* (pp. 299–316). Newbury Park, CA: Sage.

Jöreskog, K. G., & Sörbom, D. (1989). *LISREL 7 a guide to the program and applications* (2nd ed.). Chicago: SPSS, Inc.

Jöreskog, K. G., & Sörbom, D. (1993). *LISREL 8 user's reference guide.* Chicago: Scientific Software International.

Jöreskog, K. G., & Yang, F. (1996). Nonlinear structural equation models: The Kenny and Judd model with interaction effects. In G. A. Marcoulides & R. E. Schumacker (Eds.), *Advances in structural equation modeling techniques* (pp. 57–88). Hillsdale, NJ: Lawrence Erlbaum Associates.

Kendall, M. G., & Stuart, A. (1958). *The advanced theory of statistics* (Vol. 1). London: Griffin.

Kenny, D. A. (1985). Quantitative methods for social psychology. In G. Lindsey & E. Aronson (Eds.), *Handbook of social psychology* (3rd ed., Vol. 1, pp. –). New York: Random House.

Kenny, D. A., & Judd, C. M. (1984). Estimating the nonlinear and interactive effects of latent variables. *Psychological Bulletin, 96,* 201–210.

Lohmöller, J. (1981). *LVPLS 1.6 latent variables path analysis with partial least squares estimation.* University of Munich: University of the Federal Armed Forces.

Lubin, A. (1961). The interpretation of significant interaction. *Educational and Psychological Measurement, 21,* 807–817.

Lubinski, D., & Humphreys, L. G. (1990). Assessing spurious "Moderator Effects": Illustrated substantively with the hypothesized ("synergistic") relation between spatial and mathematical ability. *Psychological Bulletin, 107,* 385–393.

McClelland, G. H., & Judd, C. M. (1993). Statistical difficulties of detecting interactions and moderator effects. *Psychological Bulletin, 114*(2), 376–390.

McDonald, R. P. (1981). The dimensionality of tests and items. *British Journal of Mathematical and Statistical Psychology, 34,* 100–117.

Ping, R. A., Jr. (1993). The effects of satisfaction and structural constraints on retailer exiting, voice, loyalty, opportunism, and neglect. *Journal of Retailing, 69,* 320–352.

Ping, R. A., Jr. (1995). A parsimonious estimating technique for interaction and quadratic latent variables. *The Journal of Marketing Research, 32,* 336–347.

Ping, R. A., Jr. (1996a). Latent variable interaction and quadratic effect estimation: A two-step technique using structural equation analysis. *Psychological Bulletin, 119,* 166–175.

Ping, R. A., Jr. (1996b). Latent variable regression: A technique for estimating interaction and quadratic coefficients. *Multivariate Behavioral Research, 31*(1), 95–120.

Rusbult, C. E., Zembrodt, I. M., & Gunn, L. K. (1982). Exit, voice, loyalty, and neglect: Responses to dissatisfaction in romantic involvement. *Journal of Personal and Social Psychology, 43,* 1230–1242.

Satorra, A., & Bentler, P. M. (1988). Scaling corrections for chi-squared statistics in covariance structure analysis. *Proceedings of the American Statistical Association,* 308–313.

Sharma, S., Durand, R. M., & Gur-Arie, O. (1981). Identification and analysis of moderator variables. *Journal of Marketing Research, 18,* 291–300.

Sharma, S., Durvasula, S., & Dillon, W. R. (1989). Some results on the behavior of alternative covariance structure estimation procedures in the presence of nonnormal data. *Journal of Marketing Research, 26,* 214–221.

Stone-Romero, E. F., Alliger, G. M., & Aguinis, H. (1994). Type II error problems in the use of moderated multiple regression for the detection of moderating effects of dichotomous variables. *Journal of Management, 20*(1), 167–178.

Tanaka, J. S. (1984). *Some results on the estimation of covariance structure models.* Unpublished doctoral dissertation, University of California, Los Angeles.

Wong, S. K., & Long, J. S. (1987). Reparameterizing nonlinear constraints in models with latent variables (Technical Report). Pullman: Washington State University.

APPENDIX
Calculated Interaction and Quadratic Variances and Covariances

The following is based on Ping (1996b) and presents corrections for unadjusted variances and covariances (e.g., SAS or SPSS values) involving interactions and quadratics.

An estimate of the variance of the latent variable ξ_X using the variance of the observed variable $X = (x_1 + x_2)/2$, where x_1 and x_2 are the observed indicators of ξ_X (i.e., $x_1 = \lambda_{X1}\xi_X + \varepsilon_{X1}$ and $x_2 = \lambda_{X2}\xi_X + \varepsilon_{X2}$), x_1 and x_2 are independent of ε_{X1} and ε_{X2} (i.e., ξ_X is independent of its measurement errors), ε_{X1} and ε_{X2} are independent of each other, and x_1 and x_2 are multivariate normal with zero means, is given by the following. Let $\Gamma_a = (\lambda_{a1} + \lambda_{a2})/2$. Then

$$\begin{aligned}
\mathrm{Var}(X) &= \mathrm{Var}[(x_1 + x_2)/2] \\
&= \mathrm{Var}[\Gamma_X\xi_X + (\varepsilon_{X1} + \varepsilon_{X2})/2] \\
&= \Gamma_X{}^2\mathrm{Var}(\xi_X) + [\mathrm{Var}(\varepsilon_{X1}) + \mathrm{Var}(\varepsilon_{X2})]/2^2 \\
&\quad (= \Gamma_X{}^2\mathrm{Var}(\xi_X) + [\mathrm{Var}(\varepsilon_{X1}) + \mathrm{Var}(\varepsilon_{X2})]/4) \\
&= \Gamma_X{}^2\mathrm{Var}(\xi_X) + \theta_X, \qquad \text{(A1)}
\end{aligned}$$

where Var(a) is the variance of a, Var(X) is the observed variance of X, and $\theta_X = [\mathrm{Var}(\varepsilon_{X1}) + \mathrm{Var}(\varepsilon_{X2})]/2^2$. As a result, an estimate of $\mathrm{Var}(\xi_X)$ is given by

$$\mathrm{Var}(\xi_X) = (\mathrm{Var}(X) - \theta_X)/\Gamma_X{}^2.$$

For $\mathrm{Cov}(\xi_X\xi_Z)$, where Cov(a,b) is the covariance of a and b,

$$\begin{aligned}
\mathrm{Cov}(X,Z) &= \mathrm{Cov}[(x_1 + x_2)/2, (z_1 + z_2)/2] \\
&= [\mathrm{Cov}(\lambda_{X1}\xi_X + \varepsilon_{X1},\lambda_{Z1}\xi_Z + \varepsilon_{Z1}) + \mathrm{Cov}(\lambda_{X1}\xi_X + \xi_{X1},\lambda_{Z2}\xi_Z + \varepsilon_{Z2}) \\
&\quad + \mathrm{Cov}(\lambda_{X2}\xi_X + \varepsilon_{X2},\lambda_{Z1}\xi_Z + \varepsilon_{Z1}) + \mathrm{Cov}(\lambda_{X2}\xi_X + \varepsilon_{X2},\lambda_{Z2}\xi_Z + \varepsilon_{Z2})]/2^2 \\
&= \mathrm{Cov}(\xi_X,\xi_Z)\Gamma_X\Gamma_Z, \qquad \text{(A2)}
\end{aligned}$$

and an estimate of $\mathrm{Cov}(\xi_X\xi_Z)$ is given by

$$\mathrm{Cov}(\xi_X,\xi_Z) = \mathrm{Cov}(X,Z)/\Gamma_X\Gamma_Z,$$

where $Z = (z_1 + z_2)/2$.

Off-diagonal terms comprised of an interaction and a linear variable that does not appear in the interaction such as $\mathrm{Cov}(\xi_V,\xi_W\xi_X)$ are estimated as follows:

$$\text{Cov}(V,WX) = \text{Cov}(\Gamma_V\xi_V + E_V\,[\Gamma_W\xi_W + E_W][\Gamma_X\xi_X + E_X]),$$

where $E_a = (\varepsilon_{a1} + \varepsilon_{a2})/2$. Hence

$$\text{Cov}(V,WX) = \text{Cov}(\xi_V,\xi_W\xi_X)\Gamma_V\Gamma_W\Gamma_X,$$

and

$$\text{Cov}(\xi_V,\xi_W\xi_X) = \text{Cov}(V,WX)/\Gamma_V\Gamma_W\Gamma_X.$$

The covariance of two interactions with no common linear variables is given by

$$\text{Cov}(VW,XZ) = \text{Cov}(V,X)\text{Cov}(W,Z) + \text{Cov}(V,Z)\text{Cov}(W,X), \qquad \text{(A3)}$$

(Kendall & Stuart, 1958), and

$$\begin{aligned}\text{Cov}(VW,XZ) &= \text{Cov}(\xi_V,\xi_X)\Gamma_V\Gamma_X\text{Cov}(\xi_W,\xi_Z)\Gamma_W\Gamma_Z \\ &\quad + \text{Cov}(\xi_V,\xi_Z)\Gamma_V\Gamma_Z\text{Cov}(\xi_W,\xi_X)\Gamma_W\Gamma_X \\ &= \text{Cov}(\xi_V\xi W,\xi_X\xi_Z)\Gamma_V\Gamma_W\Gamma_X\Gamma_Z,\end{aligned}$$

by Equality (A2). An estimate of $\text{Cov}(\xi_V\xi_W,\xi_X\xi_Z)$ is therefore given by

$$\text{Cov}(\xi_V\xi_W,\xi_X\xi_Z) = \text{Cov}(VW,XZ)/\Gamma_V\Gamma_W\Gamma_X\Gamma_{Z\cdot 901t0b0s10.0v1P} \qquad \text{(A4)}$$

By Equality (A4) the covariance of two quadratics such as $\text{Cov}(\xi_X\xi_X,\xi_Z\xi_Z)$ is

$$\text{Cov}(\xi_X\xi_X,\xi_Z\xi_Z) = \text{Cov}(XX,ZZ)/\Gamma_X^2\Gamma_Z^2.$$

For the variance of an interaction

$$\begin{aligned}\text{Var}(XZ) &= \text{Cov}(XZ,XZ) \\ &= \text{Var}(X)\text{Var}(Z) + \text{Cov}(X,Z)^2,\end{aligned}$$

using Equality (A3). Hence

$$\begin{aligned}\text{Var}(XZ) &= [\Gamma_X^2\text{Var}(\xi_X) + \theta_X][\Gamma_Z^2\text{Var}(\xi_Z) + \theta_Z] + [\text{Cov}(\xi_X,\xi_Z)\Gamma_X\Gamma_Z]^2 \\ &= \text{Cov}(\xi_X\xi_Z,\xi_X\xi_Z)\Gamma_X\Gamma_Z\Gamma_X\Gamma_Z + \text{Var}(\xi_X)\Gamma_X^2\theta_Z \\ &\quad + \text{Var}(\xi_Z)\Gamma_Z^2\theta_X + \theta_X\theta_Z,\end{aligned}$$

using (A1) and (A2), and

$$\begin{aligned}\mathrm{Var}(\xi_X\xi_Z) &= (\mathrm{Var}(XZ) - \mathrm{Var}(\xi_X)\Gamma_X^2\theta_Z - \mathrm{Var}(\xi_Z)\Gamma_Z^2\theta_X - \theta_X\theta_Z)/\Gamma_X^2\Gamma_Z^2 \\ &= (\mathrm{Var}(XZ) - \theta_Z\mathrm{Var}(X) - \theta_X\mathrm{Var}(Z) + \theta_X\theta_Z)/\Gamma_X^2\Gamma_Z^2.\end{aligned}$$

The corrected estimate of a quadratic such as Var(XX) is similar:

$$\begin{aligned}\mathrm{Var}(XX) &= 2\mathrm{Var}(X)^2 \\ &= 2[\Gamma_X^2\mathrm{Var}(\xi_X) + \theta_X]^2 \\ &= \mathrm{Var}(\xi_X\xi_X)\Gamma_X^4 + 4\mathrm{Var}(\xi_X)\Gamma_X^2\theta_X + 2\theta_X^2,\end{aligned}$$

by Equality (A3), and

$$\begin{aligned}\mathrm{Var}(\xi_X\xi_X) &= (\mathrm{Var}(XX) - 4\mathrm{Var}(\xi_X)\Gamma_X^2\theta_X - 2\theta_X^2)/\Gamma_X^4 \\ &= (\mathrm{Var}(XX) - 4\mathrm{Var}(X)\theta_X + 2\theta_X^2)/\Gamma_X^4.\end{aligned}$$

For the covariance of a quadratic and an interaction that has a common linear variable such as Cov(XX,XZ),

$$\begin{aligned}\mathrm{Cov}(XX,XZ) &= 2\mathrm{Var}(X)\mathrm{Cov}(X,Z) \\ &= 2[\Gamma_X^2\mathrm{Var}(\xi_X) + \theta_X]\mathrm{Cov}(\xi_X,\xi_Z)\Gamma_X\Gamma_Z \\ &= \mathrm{Cov}(\xi_X\xi_X,\xi_X\xi_Z)\Gamma_X^2\Gamma_X\Gamma_Z + 2\mathrm{Cov}(\xi_X,\xi_Z)\Gamma_X\Gamma_Z\theta_X,\end{aligned}$$

by Equalities (A2) and (A3), and

$$\begin{aligned}\mathrm{Cov}(\xi_X\xi_X,\xi_X\xi_Z) &= (\mathrm{Cov}(XX,XZ) - 2\mathrm{Cov}(\xi_X,\xi_Z)\Gamma_X\Gamma_Z\Gamma_X)/\Gamma_X^3\Gamma_Z \\ &= (\mathrm{Cov}(XX,XZ) - 2\mathrm{Cov}(X,Z)\theta_X)/\Gamma_X^3\Gamma_Z.\end{aligned}$$

For a combination of interactions with a common linear variable such as Cov(VW,VZ)

$$\begin{aligned}\mathrm{Cov}(VW,VZ) &= \mathrm{Var}(V)\mathrm{Cov}(W,Z) + \mathrm{Cov}(V,Z)\mathrm{Cov}(W,V) \\ &= [\Gamma_V^2\mathrm{Var}(\xi_V) + \theta_V]\mathrm{Cov}(\xi_W,\xi_Z)\Gamma_W\Gamma_Z \\ &\quad + \mathrm{Cov}(\xi_V,\xi_Z)\Gamma_V\Gamma_Z\mathrm{Cov}(\xi_W,\xi_Z)\Gamma_W\Gamma_V \\ &= \mathrm{Cov}(\xi_V\xi_W,\xi_V\xi_Z)\Gamma_V^2\Gamma_W\Gamma_Z + \mathrm{Cov}(\xi_W,\xi_Z)\Gamma_W\Gamma_Z\theta_V,\end{aligned}$$

by Equalities (A2) and (A3), and

$$\mathrm{Cov}(\xi_V\xi_W,\xi_V\xi_Z) = (\mathrm{Cov}(VW,VZ) - \mathrm{Cov}(W,Z)\theta_V)/\Gamma_V^2\Gamma_W\Gamma_Z.$$

By induction these estimates can be generalized to latent variables with an arbitrary number of indicators, for example, $V = (v_1 + v_2 + \ldots + v_p)/p$, where v_i are the observed indicators of ξ_V (i.e., $v_i = \lambda_{vi}\xi_V + \varepsilon_{vi}$). θ_V is given by $\theta_V = [\mathrm{Var}(\varepsilon_{v1}) + \mathrm{Var}(\varepsilon_{v2}) + \ldots + \mathrm{Var}(\varepsilon_{vp})]/p^2$.

5

Estimating Interaction and Nonlinear Effects With SAS

Phillip K. Wood
Darin J. Erickson
University of Missouri-Columbia

MODELING INTERACTION AND QUADRATIC EFFECTS FOR LATENT VARIABLE MODELS

Polynomial regression techniques have often been used to describe the functional relationship of quadratic or interaction effects in addition to main effects of variables. Although there is a clear conceptual utility in estimating such quadratic and interaction effects, such regression models have been shown to produce biased and/or inconsistent estimates if the predictor variables contain significant amounts of measurement error (Busemeyer & Jones, 1983).

Kenny and Judd (1984) proposed an alternate model for the estimation of interaction and quadratic effects in latent variables as a remedy to this problem. (This chapter hereafter uses the term *polynomial latent variable model* to refer to both the quadratic and cross-product models proposed by Kenny and Judd.) This approach, in its original form, involves calculation of appropriate product variables, which, under the assumed measurement model and the assumption that error and exogenous factor variances are normally distributed, allows specification of the desired quadratic or product latent variable(s). Estimation of such effects, however, also requires the specification of several additional latent variables in order to model cross-product effects of factors and uniquenesses assumed under the model. Such cross-product effects, however, are often merely "nuisance" parameters in that they are unrelated to the conceptual hypotheses of interest.

Estimation Difficulties Associated With Polynomial Models

With the advent of structural equation modeling software that allows specification of nonlinear constraints and accompanying increases in computing speed, researchers have been increasingly interested in applying and evaluating such nonlinear effects. Although the elegance of the Kenny and Judd (1984) model is attractive, writing the structural equation program setups for such models is tedious and prone to human programming error (Jöreskog & Yang, 1996; Ping 1995; 1996). In addition, Ping (1995) noted that latent variable models seem especially prone to convergence problems and infeasible solutions. Remedies to such infeasible solutions in traditional structural models involve respecification of the original model. For example, McDonald (1980) discussed how observed negative error variances may be dealt with by respecifying the error variance as unity with a free path from the error variance to the manifest variable. Such an approach is not often employed in nonlinear factor models, however, because such respecification involves an entirely different set of constraints in the model, thereby doubling the program tedium necessary to specify the model. Even if a researcher desired to expend the effort to program this new set of constraints, there is no guarantee that any particular respecification will converge on a correct solution, leaving the researcher with a large menu of tedious programs to write, most of which may not achieve the desired result of representing a feasible solution.

Interpretational Difficulties With Polynomial Models

Given the sometimes formidable challenges involved in securing interpretable solutions for polynomial latent variable models, interpretational difficulties associated with these models are often underdiscussed. Although these are discussed in more detail later, it is helpful at this point to briefly note some criticisms that a reasonable skeptic might raise in response to any proposed polynomial latent variable model:

1. Given that published examples of the Kenny and Judd (1984) model assume a simple factor structure, are proposed polynomial effects due to model misspecification? For example, do any proposed polynomial effects still obtain when complex factor models for the data are considered (such as autocorrelated error covariances over time or factorially complex items).
2. To what extent do different estimation techniques result in different conclusions regarding a proposed polynomial effect?
3. Can a potentially significant polynomial effect be eliminated by elimination of a relatively small number of influential observations?

This chapter seeks to simplify the estimation of standard polynomial factor models and to facilitate the estimation of such models under different factor structures and estimation techniques through use of an easy-to-use SAS macro that takes, as input, a PROC CALIS model for estimation of the traditional measurement model and that produces a completely parameterized polynomial model in either the original Kenny and Judd (1984) approach or in an equivalent, but reduced form involving nonlinear constraints. The way in which the nonlinear constraints are derived in the macro is explained by reference to an original example taken from Kenny and Judd's article. We also take this opportunity to share our procedures that have worked well for us in securing proper solutions for polynomial factor models. Finally, selected analyses are presented from a real-world data set involving prospective assessment of problematic alcohol use and subjective expectations regarding alcohol's effects. Although the real-world nature of this data set precludes knowledge of the true model that underlies the data, this example illustrates how complex factor models may be programmed using the macro and also serves to illustrate a general observation in practice on our part that generalized least squares (GLS) and maximum likelihood (ML) solutions produce polynomial effects similar to those obtained under analogous polynomial regression models, whereas weighted least squares (WLS) solutions appear to produce quite different results.

On the Necessity of Start Values for Estimation of Polynomial Factor Models

At this point it seems unclear as to whether the convergence and infeasible solution difficulties are due to the specification of additional variables, as suggested by Ping (1995), or whether the optimization terrain for polynomial latent variable models contains several local minima that often preclude arrival at the true local minimum value of the fit function. In either situation, it seems reasonable to believe, however, that the algorithms used by structural equation programs for generating start values are less than optimal for polynomial applications. Such algorithms often do not incorporate the additional information about relationships of factor loadings and uniquenesses of the linear measurement model with factor variances, loadings, and error variances associated with product or interaction terms. It seems reasonable, therefore, to believe that a latent variable model with polynomial effects has a greater chance of converging if start values taken from the linear measurement model were used to specify start values for all elements of the corresponding polynomial model. Unfortunately, using information from the final estimates of the linear model to generate feasible start values for polynomial effects involves a large amount of hand calculation and increases the programming tedium yet again by another order of magnitude.

Outline

Original and Reduced Models for Polynomial Effects. After a short review of polynomial models employing latent variables, the present chapter introduces a small SAS program that takes as input the PROC CALIS program setup for a structural model that estimates only linear effects. This program then automatically generates the Calis program setup for a Kenny and Judd–style model for a desired latent variable model of interactive and/or quadratic effects. The program allows researchers to generate the interactive or quadratic effects under common model respecifications, such as those described by McDonald (1980), in order to remediate improper estimated solutions. In addition, the program will also automatically generate a more parsimonious, but mathematically equivalent structural model for interaction effects that does not contain the cross-product "nuisance effects" of the original Kenny and Judd formulation. The rationale for such an approach is discussed using one of Kenny and Judd's original simulated data examples. Use of the GENINT program to produce the original Kenny and Judd model as well as the reduced model is presented for an example taken from the original Kenny and Judd (1984) article.

Analysis of Real-World Data. In practice, fitting interactive and quadratic factor models using real-world data has often proved more difficult than presentations using generated data would suggest. To show how the SAS program can be used in such situations, data from a prospective study are used to investigate a prior research claim that heavy alcohol use is a function of prior alcohol expectancies, prior heavy alcohol use, and the interaction of prior alcohol expectancies with prior heavy alcohol use. Three approaches to testing for interaction effects are compared: a traditional multiple regression model that includes an interaction variable, an interactive model that assumes a simple factor structure in the exogenous latent variables, and an interactive model that assumes a complex factor structure reflecting serial correlation over time. Models involving interactive latent variables for the real-world data set are estimated and compared under the GLS criterion originally recommended by Kenny and Judd (1984), as well as the ML and WLS criteria as suggested in more recent articles.

PRIOR METHODS FOR QUADRATIC AND INTERACTIVE EFFECTS

Before proceeding, it is appropriate to briefly discuss some of the recent work on interactive and quadratic effects in latent variables. Recently, Bollen (1995) proposed a two-stage least squares approach to the estimation of nonlinear effects that makes no assumptions regarding the underlying distribution of unobserved variables. Although the use of limited-information estimators is an attractive and less restrictive alternative to the factor model of Kenny and Judd (1984), a comparison of such an approach with the factor models proposed by Kenny and Judd is beyond the scope of this chapter.

Several variants of the Kenny and Judd (1984) model have been proposed that are less laborious than the original Kenny and Judd model. Specifically, Jaccard and Wan (1995) specified a slight modification of the original Kenny and Judd model in that the interaction model they described imposes the correct nonlinear constraints on interaction factor loadings and factor variances, but allows covariances between manifest product variables to be freely estimated. Although the example analysis they presented in the appendix to their article contains nonlinear constraints similar to those calculated by the proposed program, their model is not strictly equivalent to either the Kenny and Judd approach or that proposed here. In an attempt to limit the tedium associated with modeling all variables, Jaccard and Wan calculated only some of the possible product variables in the model. In addition (and because of the omission of some product variables), error variances and covariances in the model were not constrained to be nonlinear functions of already estimated parameters. Although one may view this example as a *reduced* version of the original Kenny and Judd model, it may also be argued that it constitutes a *relaxed* version of the Kenny and Judd approach because it does not specifically constrain error variances and covariances. To the extent that measurement error and/or cross-product factors do not meet the normality assumption assumed by Kenny and Judd, such models may better fit the data. (This topic is revisited in the discussion of Jöreskog & Yang's, 1996, chapter later.)

Ping (1995) proposed two simple alternatives to the original Kenny and Judd (1984) model that are less tedious and appear to perform well in simulated data. One approach, described as a "two-step" estimation technique, involves solving the linear measurement model as a way of recovering factor loadings, and then using these estimates as fixed values in the larger polynomial model. A second, more parsimonious model is developed based on the fact that the product or quadratic expressions of the sum of variables in the linear measurement model should have a known relationship to the product or quadratic latent variable of interest. Such an approach has the advantage that only one additional manifest and latent variable is needed to estimate the desired interactive or quadratic effect. Unfortunately, the approach also assumes that the manifest variables load on only one latent variable (although Ping also proposed an indirect two-step estimation technique for complex factor models). Ping's refinements of the original Kenny and Judd approach are simple, elegant, and may constitute a more preferable avenue for many researchers, especially given the fact that the ratio of observed variables and estimated parameters to sample size may be unfavorable for many models. The advantages of using the program proposed here are that convergence rates and the likelihood of a proper solution appear to be quite good using the proposed computer program and use of the program allows for simultaneous estimation of cross-loadings and correlated errors without using Ping's proposed two-step approach.

Finally, Jöreskog and Yang (1996) proposed that researchers analyze an adjoined matrix that incorporates mean-level information in the structural model of polynomial effects. Like Jaccard and Wan (1995), Jöreskog and Yang have also noted that the advent of the ability to program nonlinear constraints in structural equation modeling programs eliminates the need to specify the cross-product parameters. As such, the restated model is an example of the reduced model proposed here for estimating interactive effects. Jöreskog and Yang provided an example respecification of the original Kenny and Judd (1984) model and their own restatement based on the adjoined variance/covariance matrix. Because they correctly noted that the mean of a product vector cannot be zero if its constituents contain a nonzero covariance, they recommended that all analyses for interactive effects be done using the adjoined variance/covariance matrix. We fail to understand why it must necessarily follow that an adjoined matrix must therefore be used, however. Multiple regression models with quadratic and interaction terms are widely used based on centered data without the need to incorporate information on the mean associated with all predictor vectors, for example. In any event, the adjoined model proposed by Jöreskog and Yang can also be implemented using the SAS program described here, but the resulting PROC CALIS program must be slightly modified.

Jöreskog and Yang's (1996) chapter is also interesting in that they considered the effects of non-normality by allowing the factor variances associated with the interaction latent variable to be free, as opposed to their constrained estimates under normal theory. They also explored whether product terms of the error variances are the product of normal variates. One of the points not explicitly discussed in the presentation, however, are the ways in which the expression of the model (particularly in reduced form) must be changed in response to such nonlinearity. In this chapter we also consider such relaxations and briefly discuss how to relax such assumptions by editing the output produced by the GENINT program.

In summary, advances on Kenny and Judd's (1984) original polynomial latent variable model have attempted to deal with difficulties encountered in practical applications of the technique. Nearly all have mentioned that writing such programs is tedious and error-prone. Although reduced expressions of the model are attractive alternatives to the original model, some of these models require the researcher to assume a simple structure for the data, others require a two-step approach that fixes some parameters while freeing others, and many individuals note convergence problems when implementing the technique. Some approaches, such as Ping's (1995) parsimonious estimation technique, are probably the avenue of choice for many researchers. For others wishing to employ the original Kenny and Judd approach, however, the SAS program outlined here produces mathematically proper solutions with a minimum of computational and programming effort. The model also allows the researcher to estimate factorially complex models in one simultaneous model.

COMPLETE AND REDUCED STRUCTURAL EQUATION MODEL REPRESENTATIONS FOR QUADRATIC AND INTERACTIVE EFFECTS

The key to a more compact notation for specification of quadratic and interactive effects results from the fact that it is possible to specify a structural equation model with only those latent variables of interest (such as squared and interaction factor models) while correctly specifying variance components of a manifest variable due to cross-product factors such as cross-product terms by including them in that part of the model associated with error variance. Similarly, covariances between manifest product variables that arise due to these cross-product factors can be expressed as an additional covariance between error variances. A simulated data example of this helps to illustrate the approach.

A Simulated Data Example. Consider the quadratic model proposed by Kenny and Judd (1984) in which two variables, A and B, are indicators of a factor, $F1$. This factor is thought to contain both linear and quadratic contributions for a criterion variable, Y. In this notation, capital letters denote latent variables, e is used to index uniquenesses, and lowercase letters are used for estimated factor loadings. The linear measurement model in equation form for this model is expressed as:

$$\begin{bmatrix} A \\ B \end{bmatrix} = \begin{bmatrix} F_1 + e_1 \\ bF_1 + e_2 \end{bmatrix} \tag{1}$$

where $F1$ represents the exogenous latent variable of interest, b represents a factor loading associated with manifest variable B, and e is used to index the error or uniqueness terms of the model. If the resulting model results in an improper solution, the model may be re-expressed by fixing the error variances to unity and estimating a free parameter with the e_1 and e_2 variables.

Accordingly, the measurement model for cross-product variables A^2, B^2, and AB (which is necessary to derive a measurement model for the quadratic latent variable, $F1F1$) is estimated by multiplying the corresponding measurement models together. That is:

$$\begin{bmatrix} A \\ B \end{bmatrix} [A\ B] = \begin{bmatrix} A^2 & AB \\ AB & B^2 \end{bmatrix} = \begin{bmatrix} F1 + e_1 \\ bF1 + e_2 \end{bmatrix} \begin{bmatrix} F1 + e_1 & bF1 + e_2 \end{bmatrix} = \begin{bmatrix} F1^2 + 2F1e_1 + e_1^2 & bF1^2 + F1e_2 + bF1e_1 + e_1e_2 \\ bF1^2 + F1e_2 + bF1e_1 + e_1e_2 & b^2F1^2 + 2bF1e_2 + e_2^2 \end{bmatrix}.$$

The measurement model for the full Kenny and Judd (1984) model for the predictor variables A, B, A^2, AB, and B^2 is, therefore,

$$\begin{bmatrix} A \\ B \\ A^2 \\ AB \\ B^2 \\ Y \end{bmatrix} = \begin{bmatrix} \mathbf{F1} + e_1 \\ \boldsymbol{b}\mathbf{F1} + e_2 \\ \mathbf{F1^2} + 2\mathrm{F}1e_1 \\ \boldsymbol{b}\mathbf{F1^2} + \mathrm{F}1e_2 + b\mathrm{F}1e_1 + e_1e_2 \\ \boldsymbol{b^2}\mathbf{F1^2} + 2b\mathrm{F}1e_2 + e_2{}^2 \\ \boldsymbol{c}\mathbf{F1} + \mathbf{dF1^2} + e_3 \end{bmatrix}.$$

In this model, all variance estimates for cross-product effects are constrained to their theoretically known values under normal theory. (These are discussed in more detail in Kenny and Judd, 1984.) $F1^2$, for example, is assumed to be equal to $2\sigma^2_{F1}\sigma^2_{F1}$. Clearly, however, the only latent variables of conceptual interest in the model are $F1$ and $F1^2$ (the terms indicated in boldtype in the equation). The other terms, which are cross-products of the latent variable with error terms, are only necessary to assure (a) that the observed variances associated with the product variables are modeled correctly, and (b) that covariances between product variables due to cross-product terms are also correctly modeled. As such, it is possible to rewrite the Kenny and Judd model with only the desired latent variables of $F1$ and $F1^2$ by specifying covariances between error terms of the manifest variables to account for the common cross-products between product variables. For example, specifying that (a) the covariance between the errors associated with A^2 and AB is equal to $2 * \mathrm{b} * \sigma^2_{F1} * \sigma^2_{e1}$ and (b) that the error variance associated with A^2, $e_1{}^2 = 2e_1{}^2 + 4 * \sigma^2_{F1} * \sigma^2_{e1}$ and that the error variance associated with AB, e_1e_2, contains similar terms to model both the variance of AB due to $F1e_1$ and $F1e_2$.

Estimation Using SAS Program. The SAS program for estimating these effects can be easily used to generate the PROC CALIS program for either the full Kenny and Judd (1984) model or the reduced model. The program requires only two types of input: a PROC CALIS program representing the basic measurement model (in this case the measurement model for variables A and B described in Equation 1 earlier) and information describing the desired targets for the interactive model and their total number. For the reduced model, only two lines of SAS code are required. That is,

```
TARGET1='f1f1';
TINDEX=1;
```

If the standard Kenny and Judd model is desired that contains all cross-product variables (which may be useful in exploring some distributional assumptions of the technique as described later), the researcher needs to specify all the cross-product terms. That is,

```
TARGET1='f1f1';
TARGET2='f1e1';
TARGET3='f1e2';
TINDEX=3;
```

After executing the program with these inputs, information regarding the structural model for the dependent variables of interest needs to be added. Because the dependent variables of interest in this case is a single variable, this consists of a single statement in the LINEQS section that states the measurement model for Y as given in the sixth row of the previous measurement model, and the addition of an error variance for this variable, $e3$, in the STD section of the program. (The complete example, as well as the resulting PROC CALIS programs are available from the gopher client at psysparc.psyc.missouri.edu.)

The question, of course, naturally arises as to why a researcher would be interested in this mathematically equivalent expression of the model as opposed to the original Kenny and Judd (1984) approach. At least two practical advantages derive from such a restatement of the model: First, if statistical software is used to represent the model in graphic form, its presentation is much less cluttered. Second, models that solve for only the relevant constraints appear to execute much more quickly and use far less computing resources than models that use the full approach of specifying all desired and "nuisance" cross-product variables.

A REAL-WORLD DATA EXAMPLE

Comparisons of Candidate Models. As mentioned previously, comparison of GLS and ML estimates and standard errors have appeared roughly comparable in examples that use generated data. Even though the GENINT program makes specification of interactive latent variable models easier, it seems worthwhile to compare the magnitude of these effects with those calculated from the more familiar multiple regression approach with cross-product terms. Given the ease with which latent variable interactive models can be generated and tested, it also seems appropriate to compare estimated parameters under GLS, ML, and WLS estimation for a real-world data set.

In addition, space considerations do not permit a full presentation of many modeling variations that can be pursued. We do, however, take a moment to discuss the general pattern of results that we encountered in our effort to understand the consonance of the observed findings under alterations to the model and data. Before continuing with the analysis, however, it is appropriate to discuss a rudimentary rationale for the substantive questions explored.

Research Context. Prior research has noted that the prediction of heavy alcohol use appears to be moderated by both prior heavy drinking and alcohol outcome expectancies (e.g., Cooper et al., 1992). In a test of this finding within a larger prospective study, Gotham, Sher, and Wood (1996) examined the prospective moderator relationships of heavy drinking, personality traits, sex, and four positive alcohol outcome expectancy scales on subsequent heavy drinking in a college-aged sample of 446 individuals. (Analyses proposed here, however, are based on 443 individuals because 3 individuals did not have complete data on some of the manifest variables of interest.) Gotham et al.'s analysis considered complete data records gathered from participants who were assessed 4 years following their freshman year and again 3 years later. See Kushner et al. (1994), Sher et al. (1991), and Gotham et al. (1996) for a further description of the sample and instruments. Gotham et al. used the traditional general linear models approach to test for the presence of interactions involving only composite manifest variables and a much larger set of predictor variables than only the alcohol expectation and heavy alcohol use constructs considered here. Within a larger model containing personality and sex effects, Gotham et al. found that subsequent heavy alcohol use was predicted significantly by prior alcohol expectancies, prior heavy alcohol use, and the interaction of prior alcohol expectancies and heavy alcohol use.

In the interests of brevity, the significant main effects for sex, interactions of sex with heavy alcohol use, and interactions of personality variables with previous heavy alcohol use are not addressed in these example analyses. The purpose of the present analysis is to examine the magnitude of structural relationships predicting heavy alcohol use from prior heavy alcohol use, alcohol expectancies, and the interaction of prior heavy alcohol use and alcohol expectancies. As discussed at the conclusion of this section, it is reasonable to explore several variants of the original data set in an effort to explore the robustness of the magnitude of proposed effects. In an earlier version of this chapter, we explored models involving simple and complex factor structures and analyses based on the original raw data, data that were rank transformed, and subsets of the data excluding abstaining participants or those who met liberal criteria for being multivariate outliers. Happily, analyses across these subsets and transformations of the data were quite similar. We therefore elected to present data based on a rank-normalized transform of the original data (using Proc Rank Normal = Blom) only because (a) ranking these measures reduced, but did not eliminate skewness in the alcohol consumption measures and (b) standalone measures of fit for these models seemed marginally better than those based on the original raw data.

Multiple Regression Models. The usual assessment of interaction effects, similar to the larger model considered by Gotham et al. (1996), is done by entering an additional variable composed of the product of the drinking and alcohol outcome expectancy composites to the model. A base linear regression model predicting subsequent heavy drinking from prior heavy drinking and composite alcohol outcome expectancies is given in the left-hand column of Table 5.1. To test for possible moderator effects, the product variable is merely entered as a separate predictor in the equation. As can be seen, the interaction variable reveals a statistically significant moderator of alcohol use based on the standardized coefficient associated with the product variable. Note also that the unstandardized and standardized effects associated with Year 4 heavy alcohol use is slightly larger under the interactive model than the original main effects model.[1] This change appears due to the slightly skewed nature of the Year 4 heavy alcohol use composite, which causes a suppressor effect in the data.

Simple Structure Measurement Model. Next, the PROC CALIS model for estimating the traditional measurement model for the data was specified. Specifically, the manifest variables for heavy alcohol use consisted of three variables: Subjects were asked to consider their drinking over the past month and to estimate the mean number of occasions per week that the individual: had five or more drinks in a single session (HVY), felt a little high or light-headed from alcohol (HI), or became drunk (DNK). The measurement model for the heavy alcohol use latent variable at Year 4 may be expressed in terms of the manifest variables as:

$$\begin{bmatrix} \text{HIY4} \\ \text{DNKY4} \\ \text{HVYY4} \end{bmatrix} = \begin{bmatrix} F_2 + e_5 \\ aF_2 + e_6 \\ bF_2 + e_7 \end{bmatrix}$$

where $F2$ represents the latent variable of heavy alcohol use, a and b denote factor loadings, and e_i indexes the uniquenesses associated with the manifest variables. A similar measurement model was used for heavy alcohol use at Year 7, using $F3$ as a designation of the Year 7 heavy alcohol use variable. Alcohol expectancies were measured by items from the expectancy question-

[1]It should be noted, however, that differences in the regression weights associated with these two regression models were more pronounced when based on raw, as opposed to rank-normalized data.

TABLE 5.1

Comparison of Estimates Predicting Year 7 Heavy Alcohol Use From Heavy Alcohol Use and Alcohol Expectancies at Year 4

		Multiple Regression		*Polynomial Factor Model Assuming Simple Structure*		
Parameter Estimated		*Main Effects Only*	*Main Effects & Interactions*	*GLS*	*ML*	*WLS*
Paths Predicting Year 7 Heavy Alcohol Use	Heavy Alc. Use Year 4	.5026[a] .4871*	.5189 .5132*	.5062 .4850*	.6255 .5643*	.6004 .5667*
	Alc. Exp. Year 4	.1258 .1212*	.1177 .1133*	.2321 .2424*	.0391 .0386	−.0268 −.0225
	Alc. Use x Alc. Exp.		−.0895 −.0794*	−.1443 −.1652*	−.0855 −.0835[b]	.0094 .0086
Variances	Heavy Alc. Use Year 4	1 .9396	1 .9396	1 .6322	1 .7762	1 .6366
	Alc. Exp. Year 4	1 .9910	1 .9910	1 .5321	1 .6497	1 .8074
	Alc. Use x Alc. Exp.		1 1.1662	1 .4431	1 .6623	1 .6753
Covariances	Alc. Use, Alc. Exp.	.5588 .5392*	.5588 .5392*	.5630 .3265*	.5612 .3975	.5535 .4016
R^2 Heavy Alc. Use		.3391	.3469	.4632	.4274	.3432

[a]First number in each cell denotes standardized parameter estimate. Second number denotes unstandardized estimate.

[b]$p < .10$.

naire that explored the degree to which individuals endorsed items related to Tension Reduction (E1Y4), Social Lubrication (E2Y4), Activity Enhancement (E3Y4), and Performance Enhancement (E4Y4). The measurement model for the four alcohol expectation manifest variables can be expressed as:

$$\begin{bmatrix} \mathrm{E1Y4} \\ \mathrm{E2Y4} \\ \mathrm{E3Y4} \\ \mathrm{E4Y4} \end{bmatrix} = \begin{bmatrix} F_1 + e_1 \\ cF_1 + e_2 \\ dF_1 + e_3 \\ eF_1 + e_4 \end{bmatrix}$$

where $F1$ is the general latent variable of alcohol expectancies, lowercase letters designate factor loadings, and e_i indexes uniquenesses associated with the manifest variable. When the structural model predicting only the linear effects of $F1$ and $F2$ was used to predict $F3$, the overall fit model was comparable to that of the interactive models discussed later. The final parameter estimates from these models were used as start values for the interactive models described later.

Estimation of the Interactive Effects. The measurement model for cross-product variables is calculated by multiplying the measurement models. That is:

$$\begin{bmatrix} \mathrm{E1Y4} \\ \mathrm{E2Y4} \\ \mathrm{E3Y4} \\ \mathrm{E4Y4} \end{bmatrix} \begin{bmatrix} \mathrm{HIY4} & \mathrm{DNKY4} & \mathrm{HVYY4} \end{bmatrix} = \begin{bmatrix} F_1 + e_1 \\ cF_1 + e_2 \\ dF_1 + e_3 \\ eF_1 + e_4 \end{bmatrix} \begin{bmatrix} F2 + e_5 & aF_2 + e_6 & bF_2 + e_7 \end{bmatrix}.$$

This yields the following 12 product variables necessary for estimating the interaction latent variable

$$\begin{bmatrix} \mathrm{E1Y4HIY4} & \mathrm{E1Y4DNKY4} & \mathrm{E1Y4HVYY4} \\ \mathrm{E2Y4HIY4} & \mathrm{E2Y4DNKY4} & \mathrm{E2Y4HVYY4} \\ \mathrm{E3Y4HIY4} & \mathrm{E3Y4DNKY4} & \mathrm{E3Y4HVYY4} \\ \mathrm{E4Y4HIY4} & \mathrm{E4Y4DNKY4} & \mathrm{E4Y4HVYY4} \end{bmatrix}$$

with their respective measurement models

$$= \begin{bmatrix} F_1F_2 + F_1e_5 + F_2e_1 + e_1e_5 & F_1F_2a + F_1e_6 + e_1F_2a + e_1e_6 & F_1F_2b + F_1e_7 + e_1F_2b + e_1e_7 \\ F_2Fc_1 + F_2e_2 + e_5F_1c + e_5e_2 & F_2aF_1c + e_2F_2a + e_6F_1c + e_6e_2 & F_2bF_1c + e_2F_2b + e_7F_1c + e_7e_2 \\ F_2F_1d + F_2e_3 + e_5F_1d + e_5e_3 & F_2aF_1d + e_3F_2a + e_6F_1d + e_6e_3 & F_2bF_1d + e_3F_2b + e_7F_1d + e_7e_3 \\ F_2F_1e + F_2e_4 + e_5F_1e + e_5e_4 & F_2aF_1e + e_4F_2a + e_6F_1e + e_6e_4 & F_2bF_1e + e_4F_2b + e_7F_1e + e_7e_4 \end{bmatrix}.$$

Variances for each of the cross-product variables are constrained to be equal to their values expected under normal theory. The equation relating the predictor variables of drinking, alcohol outcome expectancies, and their interaction in explaining subsequent heavy drinking is given by:

$$[F_3] = [\text{expect}F_1 + \text{blitz } F_2 + \text{interact}F_1F_2 + d_1].$$

At this point, however, it should be noted that the specification of interactive variables in an autoregressive study such as that proposed here probably does not conform to the model's assumptions. As Jöreskog and Yang (1996) noted, specifying a dependent variable as a function of interactive or quadratic effects implies that the distribution of that variable is not normal. It seems curious, then, to assume that the latent construct of heavy alcohol use is normally distributed at Year 4, but not at Year 7. Although this issue is taken up in a later section, this violation merits some concern, given that the researcher is not in a position to know whether the latent variables, manifest variable uniquenesses, or some combination of the two are nonnormally distributed.

Model Specification Using the SAS Program. As can be seen in the previous discussion, complete specification of the full Kenny and Judd (1984) model involves the computation of several cross-product terms. Use of the SAS program to generate a reduced model for interactive effects is quite straightforward, however. Again, two pieces of information are needed: the PROC CALIS program setup corresponding to the base linear measurement model, and a specification of the desired targets and the total number of targets. In this case only two statements are needed:

```
TARGET1='f1f2';
TINDEX=1;
```

The resulting SAS program was then used to investigate whether interactive effects were present. In this case, it was necessary to add four lines to the LINEQS section of the program because the dependent variable of interest was a latent variable. Three of these lines specified the measurement model for the three Year 7 heavy alcohol use manifest variables, and one line specified the measurement model for the latent variable of Year 7 heavy alcohol use. Similarly, four additions were necessary to the STD section: Three error variances associated with the manifest variables and a disturbance term associated with the Year 7 heavy drinking latent variable were added.

When this was done, GLS, ML, and WLS solutions all converged quickly and to a proper solution. Selected parameters from the GLS, ML, and WLS models are given in the three right-hand columns of Table 5.1. From this, we can see that there is substantial disagreement between the GLS, ML, and

TABLE 5.2
Selected Fit Measures for Full Factor Model
Including Interactive Effects: ML Solution

	Factor Model	
Fit Index	*Simple* (*df* = 224)	*Complex* (*df* = 231)
Model χ^2	761.6963	684.1800
AIC	303.6963	242.1800
RMSEA	.0725	.0689
Estimate (90% CI)	(.0669–.0783)	(.0630–.0748)
SBC	−633.7311	−662.4989
CFI	.9449	.9521
NNFI	.9444	.9499
NFI	.9231	.9309

WLS solutions as to the statistical significance and direction of the interaction. Although the GLS and ML solutions at least agreed with the direction of the estimate from the regression model, only in the ML solution was the estimate of the interaction effect statistically significant. The differences in estimates across ML and GLS for this real-world data set are quite different than Jaccard and Wan's (1995) conclusions using simulated data. WLS estimates of the interaction were not statistically significant and in the opposite direction from estimates found in the other models. Measures of fit for the ML solution are given in the left-hand column of Table 5.2. Although the fit measures for these data are somewhat lower than for a model of linear association, the NNFI and NFI values would be acceptable to many researchers. In contrast, the RMSEA value seems a bit high to conclude that the model is a good fit to the observed data. (Fit measures shown here are for the ML solution, although fit measures for the GLS solution were calculated and found analogous to the ML results except that many were, of course, lower than their ML counterparts.)

Factorially Complex Measurement Models. Anderson and Gerbing (1988) and others have stressed the need to first fit the basic measurement model for a given data set before proceeding to a structural analysis. Even though in this case, the fit of the model could be viewed as acceptable, it seemed reasonable to explore whether modifications of the model would result in a different pattern of results. Three conceptual reasons are relevant:

1. Such effects may constitute a class of alternative explanations for a proposed nonlinear effect (if it were found, e.g., that the more complex models no longer contained significant estimates associated with the moderator effect).

2. Assuming a simple factor structure could bias estimates of covariances between predictor factors.
3. Misspecification of the measurement model may be one reason that the variances of product variables involving either error or factor variances do not conform to the normal theory values predicted under the Kenny and Judd (1984) model.

In order to improve the measurement model with interactive effects, the measurement model for linear effects was re-examined on conceptual and empirical grounds. Specifically, it seemed reasonable to assume that the manifest variables associated with heavy alcohol use would demonstrate serial correlation over time. The measurement model for the heavy alcohol use latent variable at Years 4 and 7 were therefore respecified as:

$$\begin{bmatrix} \text{HIY4} \\ \text{DNKY4} \\ \text{HVYY4} \end{bmatrix} = \begin{bmatrix} F_2 + hF_4 + e_5 \\ aF_2 + iF_5 + e_6 \\ bF_2 + jF_6 + e_7 \end{bmatrix}$$

and

$$\begin{bmatrix} \text{HIY5} \\ \text{DNKY5} \\ \text{HVYY5} \end{bmatrix} = \begin{bmatrix} F_3 + kF_4 + e_8 \\ fF_3 + lF_5 + e_9 \\ gF_3 + mF_6 + e_{10} \end{bmatrix}.$$

where $F2$ and $F3$ represent the latent variable of problematic alcohol use in Years 4 and 7, respectively, and $F4$, $F5$, and $F6$ represent method factors that account for serial correlation in the manifest variables over time. It is worth noting that all of the serial autocorrelation loadings were statistically significant under GLS and ML analyses of the linear measurement model and these associations remained significant under the interactive models described later as well.

Changes were also made to the four alcohol outcome expectancies. Examination of the Lagrange multipliers for the linear measurement model revealed significant unmodeled covariation between the Activity Enhancement and Performance Enhancement scales not explained by the general expectancy variables. Accordingly, this association was modeled as an additional method factor with two indicators. The measurement model for the four alcohol expectation manifest variables can be expressed as:

$$\begin{bmatrix} \text{E1Y4} \\ \text{E2Y4} \\ \text{E3Y4} \\ \text{E4Y4} \end{bmatrix} = \begin{bmatrix} F_1 + e_1 \\ cF_1 + e_2 \\ dF_1 + nF_7 + e_3 \\ eF_1 + oF_7 + e_4 \end{bmatrix}$$

where $F1$ is, as before, the general latent variable of alcohol expectancies, and $F7$ is the additional method factor associated with the third and fourth expectancy scales. Fit indices for this revised linear measurement model are given on the two right-hand columns of Table 5.2. Although the increment in fit between the simple and complex measurement models is significant, examination of overall fit statistics for the model reveals only a modest improvement in an already high degree of fit to the data. Although further changes to the model were suggested by the Lagrange Multiplier tests, further refinement of the model was not undertaken because some of the suggested additions did not make conceptual sense and the overall fit of the model seemed to be adequate for the present purposes. The right-hand column of Table 5.2 shows the improvement in overall fit of the complex measurement model. Although the gains in model fit are modest, the chi-square difference test between the simple and complex models is statistically significant.

The structural models for cross-products variables that are used as indicators of the interactive latent variable are then calculated by multiplying the appropriate measurement equations. That is:

$$\begin{bmatrix} \text{E1Y4} \\ \text{E2Y4} \\ \text{E3Y4} \\ \text{E4Y4} \end{bmatrix} [\text{HIY4 DNKY4 HVYY4}] = \begin{bmatrix} \text{F}_1 + e_1 \\ c\text{F}_1 + e_2 \\ d\text{F}_1 + n\text{F}_7 + e_3 \\ e\text{F}_1 + o\text{F}_7 + e_4 \end{bmatrix} [F2 + hF_4 + e_5 \; aF_2 + iF_5 + e_6 \; bF_2 + jF_6 + e_7].$$

The resulting measurement models associated with the product variables are somewhat large and cumbersome. For example, cross-product terms for the four expectancy variables when multiplied by HIY4 are:

$$F_2\text{F}_1 + h\text{F}_1F_4 + e_5\text{F}_1 + F_2e_1 + he_1F_4 + e_1e_5,$$
$$cF_2\text{F}_1 + hF_4c\text{F}_1 + ce_5\text{F}_1 + F_2e_2 + he_2F_4 + e_2e_5,$$
$$dF_2\text{F}_1 + dhF_4\text{F}_1 + de_5\text{F}_1 + nF_2\text{F}_7 + hn\text{F}_7F_4 + ne_5\text{F}_7 + F_2e_3 + he_3F_4 + e_3e_5,$$
$$eF_2\text{F}_1 + ehF_4\text{F}_1 + ee_5\text{F}_1 + oF_2\text{F}_7 + ho\text{F}_7F_4 + oe_5\text{F}_7 + F_2e_4 + he_4F_4 + e_4e_5.$$

Cross-product terms of the four expectancy measures with DNKY4 and HVYY4 were calculated in an analogous manner.

Estimation Using SAS Program. Generation of the SAS program proceeded as described earlier for the simple-structure interaction model, except that the corresponding measurement models from the factorially complex linear model were used. As an additional check on the accuracy of estimation, this model was run twice, once specifying only the targets of interest ($F1F2$), and again, specifying all cross-product variables as described by Kenny and Judd (1984). Both programs converged to the same parameter estimates and fit function values.

Figure 5.1 shows the structural model and unstandardized coefficients associated with this more complex model under GLS estimation. Table 5.3 shows selected parameters for the latent variables in the model across estimation techniques. Use of a more complex measurement model for the data resulted in only minor changes in parameter estimates predicting the latent variable of Year 7 Heavy Alcohol Use. In this model, estimates of the interaction effect are significant for both the ML and GLS solutions, however this finding is probably only a marginal deviation in the estimated effect for the ML solution. Estimates of the interaction from the WLS solution continue to be discrepant from that obtained via multiple regression, GLS, and ML.

SUPPLEMENTAL ANALYSES

The analyses presented here represent only a sample of the variety of models that have been explored with these data. Although the GENINT program is useful in producing convergent solutions for factorially simple and complex

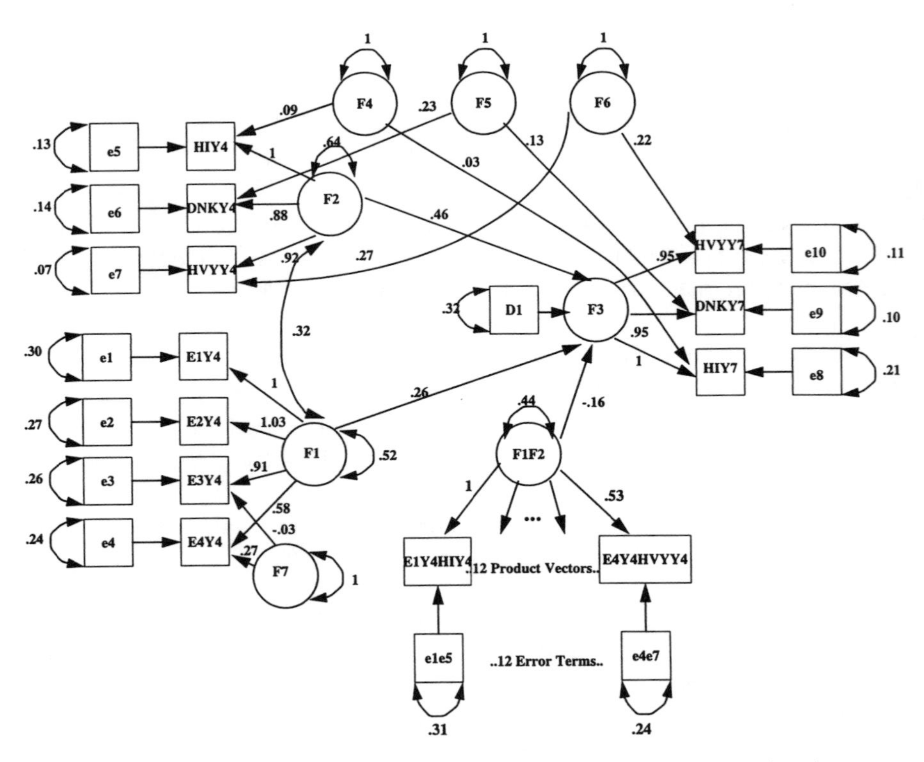

FIG. 5.1. Structural model for Year 7 Problematic Alcohol Use ($F3$) based on linear and moderator effects of Year 4 Problematic Alcohol Use ($F2$) and Year 4 Alcohol Expectations ($F1$) (coefficients in the figure are for unstandardized GLS estimates).

TABLE 5.3
Polynomial Factor Models Assuming Serial Correlation in Manifest Alcohol Use Variables and Complex Factor Structure for Alcohol Expectancies

Parameter Estimated		*Method*		
		GLS	*ML*	*WLS*
Paths Predicting Year 7 Heavy Alcohol Use	Heavy Alc. Use Year 4	.4855[a] .4651*	.5969 .5388*	.6134 .5941*
	Alc. Exp. Year 4	.2476 .2633*	.0525 .0523	.0323 .0276
	Alc. Use × Alc. Exp.	−.1399 −.1620*	−.0951 −.0941*	.0054 .0050
Variances	Heavy Alc. Use Year 4	1 .6422	1 .7733	1 .6405
	Alc. Exp. Year 4	1 .5214	1 .6327	1 .8223
	Heavy Alc. Use × Alc. Exp.	1 .4395	1 .6434	1 .6880
Covariances	Alc. Use, Alc. Exp.	.5587 .3233*	.5612 .3925	.5535 .4017
R^2 Year 7 Heavy Alc. Use		.4509	.4032	.3992

[a]First number in each cell denotes standardized parameter estimate. Second number denotes unstandardized estimate.
*$p < .05$.

models, the utility of the program is perhaps most clear when one wishes to consider the several possible models that appear reasonable. Although space prohibits description of these models in detail, the general characteristics of the analyses are discussed.

Distributional Violations. The possibility that distributional violations may produce spurious interaction effects under multiple regression or latent variable approaches is often underdiscussed. This is especially problematic for interaction models because standard software such as LISCOMP cannot be used in such situations. One method we used to explore this possibility was to rank normalize the data. For this data set, such a transform did not appreciably affect the pattern of statistical significance in linear and interactive effects and produced somewhat better measures of model fit.

Specification Errors. It is possible that other polynomial effects exist in the data besides those suggested by theory. To investigate this possibility, quadratic effects associated with Heavy Alcohol Use and Alcohol Expectancies were also estimated. This was done by merely including the targets $F1F1$ and $F2F2$ to the target list in the GENINT program and including $F1F1$ and $F2F2$ into the structural model for Year 7 Alcohol Use. These analyses also converged, but had generally poorer overall fits and failed to yield any statistically significant effect for the quadratic terms under any estimation technique for either the raw or ranked data. Given that a second expectancy factor was modeled in these data, the question as to whether this factor also contained an interaction with the latent variable of Heavy Alcohol Use, the interaction of Alcohol Use with the second expectancy factor was also calculated and used to predict subsequent Alcohol Use. This parameter also failed to explain a statistically significant amount of variability in the criterion under all methods of estimation.

Effects of Outliers. It is possible that a significant interaction effect could be due merely to the presence of a few atypical but influential data points in a study. To examine this possibility, the factorially complex interaction model was rerun excluding 10% of the individuals in the study who were highest in multivariate non-normality (Bollen & Arminger, 1991). These models were based on both raw and ranked version of the restricted sample. These analyses resulted in statistical significance for the interaction parameter across both GLS and ML.

Non-Normality. As mentioned previously, one of the restrictive assumptions of the Kenny and Judd (1984) model is its assumption that all uniquenesses and factors associated with exogenous latent variables are normally distributed. As Kenny and Judd noted in their original article, it is not possible to relax all of the normality assumptions in the model and obtain an identified solution. It is possible, however, to relax some of the normality assumptions associated with selected parts of the model. When the normality assumption is to be relaxed, it is important to consider whether the researcher believes that the variable of interest is distributed as a non-normal, but symmetric distribution, or whether no assumption regarding symmetry needs to be made. If the variable(s) of interest may be assumed to be symmetric, covariances need not be estimated between cross-product terms and their component variables. If, however, the variable(s) of interest are thought to be nonsymmetric, covariances between cross-product terms and their components must also be included in the model (as was seen, e.g., in the multiple regression analyses described that included the product term of the two composites). The requirement to include such covariances rapidly results in nonidentified interactive models.

As an example of this, the complex measurement model based on ranked data was rerun, except that the normality of the Heavy Alcohol Use latent variable was not assumed. To do this, a full specification of all cross-product terms was done in the program, and constraints on all product variables associated with the Heavy Alcohol Use latent variable were removed. In addition, covariances between the Year 4 Heavy Alcohol Use and Year 4 Alcohol Expectancy latent variables with the interaction term were estimated. The results yielded a small but statistically significant improvement in the model chi-square difference test in both the GLS and ML solutions (χ^2 = 576.1239 and 659.6696 for the GLS and ML solutions, respectively, with 213 degrees of freedom). Parameter estimates were not appreciably different for this solution and no changes in statistical significance were noted. Correlations of the Heavy Alcohol Use and Alcohol Expectancy variables with the interaction latent variable were significant, but slight (*r*s = .11 to .19 across both ML and GLS solutions). It should be noted that these correlations were much higher when the normality assumption was relaxed and the model rerun based on raw data. For example, when a GLS solution was estimated, $r = .30$ between Alcohol Expectancies and the latent variable product and $r = .45$ between Heavy Alcohol Use and the latent variable product.

Attempts, however, to further relax the normality assumption of the uniquenesses associated with the manifest variables resulted in an improper solution under both ML and GLS. Although it would be possible to explore this issue further by restating the model so that error variances could not assume negative values, it was decided to not pursue the model further.

Other Approaches to Estimation of Nonlinear Effects. With minor modifications, the GENINT program can also be used to estimate nonlinear effects according to the model outlined by Jöreskog and Yang (1996), who proposed an analysis based on the adjoined matrix taking into account mean-level information. The GENINT program can be used for such data as well, provided that the researcher adds a dummy factor to the linear measurement model to convey mean-level information that does not possess a variance. Although this extension of the model is possible using the program, we elected not to present these analyses in the interests of space.

DISCUSSION

Even though the real-world analyses presented deal with only one study, it is hoped that the variety and extent of the models considered points to the utility of the GENINT program in fitting interactive and quadratic effects. Although a comparison of computer resources necessary to fit latent variable interaction models was not presented here, use of the restricted model that employed nonlinear constraints for cross-product effects results in substantial savings. The flexibility and ease of use of the model will also hopefully encourage researchers to consider more realistic measurement models for their data.

Even though the analysis presented here concerns only one data set, some recommendations regarding practice for fitting such latent variable models with interactive effects also appear appropriate. First, researchers who wish to investigate interactive or quadratic effects at the latent variable level should pay particular attention to the linear measurement model before fitting interactive effects. Given that the GENINT program can successfully model interactive effects with factorially complex effects, it seems reasonable to explore this relatively neglected aspect of model construction involving product or quadratic effects. Second, although rank normalizing the data in order to reduce the skew of manifest variables does not appear to improve or alter fit statistics for estimates of linear effects, it appears that the fit of interactive models was substantially improved. Given that standard remedies for discontinuous scaling are not possible in interactive effects, perhaps normalizing techniques should be explored in future work as either a way to secure more consistent solutions for the data or as ancillary analyses to judge the robustness of a proposed effect. Use of more accurate measurement models is of particular value in applications of the Kenny and Judd (1984) approach; because it is a full-information estimation technique, specification errors in the base measurement model may easily introduce bias in other parameter estimates (Bollen, 1995).

A short word, however, is in order concerning the practical benefit of factor models such as those proposed here relative to simpler, more easily understood alternatives. Recall that the original purpose of the factor model for such data is to uncover the structural relationships between predictors and a criterion after adjusting for the effects of possibly differential measurement error in the exogenous variables. The increase in unique variability in the criterion accounted for by the interaction appears to be slight, however (recall, e.g., that the normalized rank-transformed standardized regression coefficient was –.09 vs. –.09 and –.14 for the ML and GLS complex factor solutions, respectively). Although other applications may recover more dramatic differences between a regression and factor approach, the value of the Kenny and Judd (1984) approach for these data appears to be more related to the fact that it can estimate interaction effects in the presence of more complex relationships among predictor variables (such as serial correlation or a complex factor structure) than can be addressed under the usual regression approach with product variables. Although it is possible to use the GENINT program to estimate a regression model corrected for attenuation due to the unreliability of the composite variables, it is unlikely that such models will be useful, owing to the bias introduced by use of internal consistency estimates in a multivariate framework.

Future work providing researchers with insight as to which techniques to use on real-world data is clearly in order. As mentioned earlier, the two-stage least squares approach of Bollen (1995) appears particularly promising in that it requires less programming effort, does not require normality, allows the researcher to explore whether other effects than simple products and quadratic effects exist, and readily allows the researcher to incorporate intercept information into the model. Similarly, Jöreskog and Yang's (1996) extension of the Kenny and Judd (1984) model also allows the researcher to incorporate intercept information into the interactive model. Clearly, even though WLS estimation has considerable aesthetic value, some caution regarding its use seems warranted, given the lack of correspondence of WLS estimates relative to the regression, GLS, and ML estimates. Certainly, additional research is needed to inform researchers as to the relative advantages of these approaches as well as their susceptibility to the effects of non-normality. Using values from the linear measurement model as fixed values (as opposed to start values) allows the researcher to specify two-step approaches to estimation that may also prove useful in developing and defending models with polynomial effects. Comparison and evaluation of these techniques using real-world data would do much to inform researchers about which techniques are most appropriate.

Regardless of the technique finally used to detect nonlinear effects, it is important for the researchers to adopt a methodological skepticism about a desired claim. Satisfactory analysis of real-world data involves the enunciation of a single structural model, determination of satisfactory model fit, and statistically significant parameters. The analyst must also "kick the tires" of a finding by considering additional analyses that a reasonable skeptic might raise as an alternative to the proposed interpretation. The presence of spurious suppressor effects due to distributional properties, the possibility of overly influential outlier observations and the biasing effects of model misspecification all argue against a single "definitive" structural model for a given study. It's hoped that the ease with which the GENINT program allows researchers to specify complex factor models and multiple quadratic and interaction effects will promote such methodological skepticism when testing for nonlinear structural relationships.

ACKNOWLEDGMENTS

This research was supported in part by Grant R01 AA07231 from the National Institute on Alcohol Abuse and Alcoholism to Kenneth J. Sher. The program and worked examples described in this chapter can be obtained at the gopher client located at psysparc.psyc.missouri.edu.

REFERENCES

Anderson, J. C., & Gerbing, D. W. (1988). Structural equation modeling in practice: A review and recommended two-step approach. *Psychological Bulletin, 103,* 453–460.

Bollen, K. A. (1995). Structural equation models that are nonlinear in latent variables: A least squares estimator. In P. V. Marsden (Ed.), *Sociological methodology* (Vol. 25, pp. 223–251). Washington, DC: American Sociological Association.

Bollen, K. A., & Arminger, G. (1991). Observational residuals in factor analysis and structural equation models. In P. V. Marsden (Ed.), *Sociological methodology* (Vol. 21, pp. 235–262). Washington, DC: American Sociological Association.

Busemeyer, J. R., & Jones, L. E. (1983). Analysis of multiplicative combination rules when the causal variables are measured with error. *Psychological Bulletin, 93,* 549–562.

Cooper, M. L., Russell, M., Skinner, J. B., Frone, M. R., & Mudar, P. (1992). Stress and alcohol use: Moderating effects of gender, coping, and alcohol expectancies. *Journal of Abnormal Psychology, 101,* 139–152.

Gotham, H. J., Sher, K. J., & Wood, P. K. (1996). *Changes in heavy drinking over the college years.* Unpublished manuscript.

Jaccard, J., & Wan, C. K. (1995). Measurement error in the analysis of interaction effects between continuous predictors using multiple regression: Multiple indicator and structural equation approaches. *Psychological Bulletin, 117,* 348–357.

Jöreskog, K. G., & Yang, F. (1996). Nonlinear structural equation models: The Kenny–Judd model with interaction effects. In G. A. Marcoulides & R. E. Schumacker (Eds.), *Advanced structural equation modeling* (pp. 57–88). Mahwah, NJ: Lawrence Erlbaum Associates.

Kenny, D. A., & Judd, C. M. (1984). Estimating the nonlinear and interactive effects of latent variables. *Psychological Bulletin, 96,* 201–210.

Kushner, M. G., Sher, K. J., & Wood, P. K. (1994). Anxiety and drinking behavior: Moderating effects of tension-reduction alcohol outcome expectancies. *Alcoholism: Clinical and Experimental Research, 18,* 852–860.

McDonald, R. P. (1980). A simple comprehensive model for the analysis of covariance structures: Some remarks on applications. *British Journal of Mathematical and Statistical Psychology, 33,* 161–183.

Ping, R. A. (1995). A parsimonious estimating technique for interaction and quadratic latent variables. *Journal of Marketing Research, 32,* 336–347.

Ping, R. A. (1996). Latent variable interaction and quadratic effect estimation: A two-step technique using structural equation analysis. *Psychological Bulletin, 119,* 166–175.

Sher, K. J., Walitzer, K. S., Wood, P. K., & Brent, E. E. (1991). Characteristics of children of alcoholics: Putative risk factors, substance use and abuse, and psychopathology. *Journal of Abnormal Psychology, 100,* 427–448.

6

Two-Stage Least Squares Estimation of Interaction Effects

Kenneth A. Bollen
Pamela Paxton
University of North Carolina at Chapel Hill

Interactions of variables occur in a variety of statistical analyses. Log-linear analyses for categorical variables routinely include solutions where the combination of values of two or more variables affect the cell counts in a contingency table. Much has been written on interactions in multiple regression models (e.g., Jaccard, Turrisi, & Wan, 1990) and empirical applications are relatively common. Yet, when we turn to *structural equation modeling* (SEM), the literature is relatively sparse. There are a few papers that propose methodologies for incorporating interactions of latent variables into equations, but few empirical examples exist. The lack of applications is not due to the failure of substantive arguments that suggest such interactions. Rather, the best known procedures for models with interactions of latent variables are technically demanding. Not only does the potential user need to be familiar with SEM, but the researcher must be familiar with programming nonlinear and linear constraints and must be comfortable with fairly large and complicated models.

The primary goal of this chapter is to provide a largely nontechnical description of an alternative technique to include interactions of latent variables in SEM. This technique avoids many of the problems associated with other methods. Specifically, our purposes are to: (a) give an overview of a *two-stage least squares* (2SLS) method to estimate interactions of latent variables, (b) provide guidance on the selection of the *instrumental variables* (IVs) that are part of the procedure, (c) contrast this method with other multiple-indicator methods,[1] and (d) compare the results of the 2SLS method with the others using both a simulation and an empirical example. Although the chapter is largely didactic, a technical version of this material that is more general and that substantiates some of the properties we describe is in Bollen (1995).

The next section gives a generic description of a model with interactions of latent variables with multiple indicators. It introduces features of the model that are common across the major methods and is used as a running example throughout the chapter. Following that, we present a section on the 2SLS method to model such interactions. Then we discuss the Kenny and Judd (1984) and related methods for handling such interactions. The next section compares the two techniques with the simulation and empirical data that Kenny and Judd provided in their original paper. Another empirical example of the 2SLS method follows. The last section makes concluding comments and contrasts the alternative methods and their properties.

INTERACTIONS OF LATENT VARIABLES

We begin with a simple single-equation model that has a dependent variable, y_1, observed without measurement error. On the right-hand side we have two latent variables and their product interaction,

$$y_1 = \alpha_{y_1} + \beta_{11} L_1 + \beta_{12} L_2 + \beta_{13} L_1 L_2 + \zeta_1 \tag{1}$$

where L_1 and L_2 are latent random variables, α_{y_1} is the intercept term, and ζ_1 is a random disturbance term with a mean of 0, having a constant variance, and without autocorrelation. The latent variables L_1 and L_2 are each measured with two indicators such that:

$$x_1 = L_1 + \delta_1 \tag{2}$$
$$x_2 = \alpha_2 + \lambda_{21} L_1 + \delta_2 \tag{3}$$
$$x_3 = L_2 + \delta_3 \tag{4}$$
$$x_4 = \alpha_4 + \lambda_{41} L_2 + \delta_4 \tag{5}$$

where α_2 and α_4 are intercept terms for Equations 3 and 5; $E(\delta_i)$ is zero; δ_i and ζ_1 are distributed independently of L_1 and L_2 and of each other; and L_1, L_2, δ_1, δ_2, δ_3, δ_4, and ζ_1 are each i.i.d. random variables.

Equation 2 shows that L_1 is scaled to have the same metric and origin as x_1. The same metric is set by the implicit coefficient of "1" for L_1 in the x_1

[1]We call the techniques "multiple indicator" to contrast them with those proposed by Busemeyer and Jones (1983), Heise (1986), and Feucht (1989) for interactions of latent variables when only a single indicator and its reliability are available for each latent variable. This is a relatively uncommon situation. Interested readers should consult the preceding references.

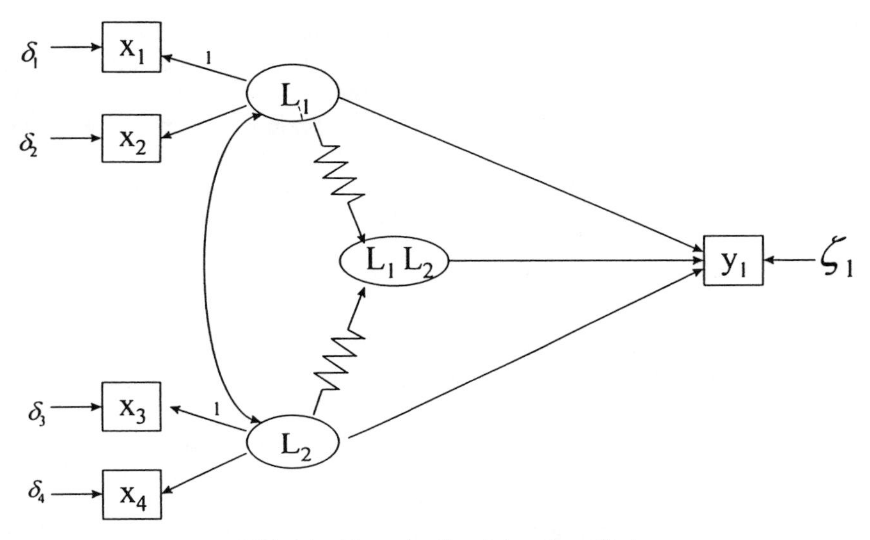

FIG. 6.1. Example of an interaction effect.

equation. A less familiar constraint is that the origin of L_1 is also set to x_1 and this is done by setting the intercept to zero in the same equation. Similarly, the metric and origin of L_2 is set to x_3. This scaling option allows us to have a scale for the latent variable that is quite similar to the scale of one of its indicators. It is not identical to it because the observed variable also is influenced by a measurement error term. One consequence is that the variance of the observed variable is greater than the variance of the latent variable. However, the observed variable used to scale the latent variable provides an approximation to the units for the latent variable when the researcher introduces the aforementioned constraints.

The path diagram for this model is in Fig. 6.1. We follow the standard path analysis convention of using circles or ovals to enclose substantive latent variables, boxes to enclose observed variables, single-headed arrows to represent direct effects, and curved two-headed arrows to show bivariate associations. There is no universal convention on representing interaction terms but we use the "saw-toothed" arrows to represent the nonlinear direct relations of L_1 and L_2 to their product interaction term, L_1L_2. The latent variable, L_1L_2, has no disturbance term because it is an exact nonlinear function of L_1 and L_2, that is, $L_1L_2 = L_1 \times L_2$.[2]

The model has two indicators per latent variable, but at this point we do not have any observable information on the latent variable interaction term, L_1L_2. Kenny and Judd (1984) recognized that if we formed the products of the indicators of the main effect latent variables, we could gain information

---

[2]The same convention is followed in Bollen (1995).

on the product of the latent variables. More specifically, take each indicator of the first latent variable and multiply it by one of the indicators of the other latent variable. Do this for each possible pair of indicators of the different latent variables. Returning to our example in Fig. 6.1, this would lead to four possible products: x_1x_3, x_1x_4, x_2x_3, and x_2x_4. Each of these products gives us information on L_1L_2. For instance, if we substitute Equation 2 for x_1 and Equation 4 for x_3 and multiply it out, we find that

$$x_1x_3 = L_1L_2 + L_1\delta_3 + L_2\delta_1 + \delta_1\delta_3. \tag{6}$$

Analogous equations for the other product indicators would show that they reflect the product latent variable term along with a composite disturbance term.

The same idea generalizes to having more than two indicators per latent variable. The multiple indicator methods for products of latent variables are alike in their need for these product variables, but they differ in their use of these terms. In the next section we discuss the 2SLS method of using this information. In a later section we review the Kenny and Judd (1984) method.

TWO-STAGE LEAST SQUARES METHOD

The 2SLS method to estimating structural equation models (SEMs) with product interactions proposed in Bollen (1995) proceeds by substituting observed variables for the latent variables in the original equation and then using instrumental variable methods to estimate the parameters. To understand the method better, consider Equations 1 to 5 that correspond to the model drawn in Fig. 6.1. Equation 1 is $x_1 = L_1 + \delta_1$ and we can easily solve this for L_1,

$$L_1 = x_1 - \delta_1. \tag{7}$$

Similarly, we can solve Equation 3 for L_2,

$$L_2 = x_3 - \delta_3. \tag{8}$$

Now, take Equations 7 and 8 and substitute them into Equation 1 in all places where L_1 and L_2 appear. This leads to

$$y_1 = \alpha_{y_1} + \beta_{11}(x_1 - \delta_1) + \beta_{12}(x_3 - \delta_3) + \beta_{13}(x_1 - \delta_1)(x_3 - \delta_3) + \zeta_1 \tag{9}$$

$$= \alpha_{y_1} + \beta_{11}x_1 + \beta_{12}x_3 + \beta_{13}x_1x_3 + u_1 \tag{10}$$

where u_1 is a composite disturbance:

$$u_1 = -\beta_{11}\delta_1 - \beta_{12}\delta_3 - \beta_{13}(x_1\delta_3 + x_3\delta_1 - \delta_1\delta_3) + \zeta_1. \tag{11}$$

Equation 10 shows that we can rewrite the latent variable model that originally was in Equation 1 into an equation of observed variables and a disturbance term. Indeed, this equation looks like the usual multiple regression equation with a product interaction term. However, the key difference between this equation and the usual multiple regression one is that the presence of measurement error leads the composite disturbance, u_1, to correlate with the observed xs on the right-hand side of Equation 10. The correlation of disturbance and explanatory variables violates a key assumption of ordinary least squares (OLS) regression. Thus, we cannot use OLS regression as an unbiased or consistent estimator of the βs.

It is at this point that the IVs and 2SLS estimator become important. IVs are observed variables that are correlated with the right-hand-side variables, but are uncorrelated with the disturbances of that equation. Analytically we can describe the 2SLS estimator as having two steps. In the first step, each of the right-hand-side variables is regressed on the IVs and the predicted value of the variable from these OLS regressions is formed. Because the predicted variables are linear combinations of the IVs, they are "uncorrelated" with the disturbance term of the original equation. The second step replaces the original right-hand-side variables with these predicted values and then uses OLS regression to estimate this modified version of Equation 10. The second-stage coefficient estimator is a consistent estimator with a known asymptotic distribution for which we can estimate standard errors and perform significance tests.

Consider the 2SLS method steps in the case of Equation 10. The first step is to find IVs that correlate with the included variables (x_1, x_3, and x_1x_3), but are uncorrelated with u_1, the disturbance. The variables included in Equation 10 are not eligible to be IVs because they correlate with the disturbance. The only other observed variables are x_2 and x_4—the other indicators of the latent variables. If these were the only IVs we would not be able to use the 2SLS estimator. This is because we need at least as many IVs as there are right-hand-side variables in Equation 10. Because we have three right-hand-side variables, we need a minimum of three IVs. It is at this point that the product indicators that we formed at the end of the prior section are utilized. These product indicators provide a pool of possibly new IVs. Besides x_1x_3, we have three product indicators: x_1x_4, x_2x_3, and x_2x_4. The first two products include the scaling indicators, either x_1 or x_3, and are correlated with some elements of u_1. Thus, they are not eligible to be IVs. However, the product x_2x_4 is not correlated with the disturbance and is a suitable IV. Thus, for Equation 10, our list of IVs is x_2, x_4, and x_2x_4.

With our IVs selected we can apply the 2SLS procedure:

1. Estimate the OLS regressions of: x_1 on x_2, x_4, and x_2x_4 and form $\hat{x}_1$; x_3 on x_2, x_4, and x_2x_4 and form $\hat{x}_3$; x_1x_3 on x_2, x_4, and x_2x_4 and form $\widehat{x_1x_3}$.
2. Estimate the OLS regression of y_1 on $\hat{x}_1$, $\hat{x}_3$, and $\widehat{x_1x_3}$.

The coefficient estimates from Step 2 are the 2SLS estimates. They are consistent estimators of the corresponding coefficients in the original Equation 1 with latent variables. Using a 2SLS procedure, the researcher also can obtain correct asymptotic standard errors for significance tests. (These standard errors are not the same as the standard errors from Step 2.). In addition, the estimator applies even when the observed variables come from non-normal distributions (Bollen, 1996).

Choosing Instrumental Variables

2SLS estimators are widely available in standard statistical packages such as SAS, SPSS, and STATA. So, implementing this procedure is relatively easy. Choosing the IVs is perhaps the most difficult part of this approach. However, in the interaction models that are most common, it is possible to give some straightforward rules to help in the selection of IVs. The preceding discussion clarified the formation and selection of IVs in the case of two latent variables, each measured with two indicators, and a product interaction. To simplify we assume that there are no correlated errors of measurement in the models. We can handle correlated errors, but it complicates our discussion. So, for didactic purposes we do not treat it here. As we showed previously, the latent variable model in Equation 1 is $y_1 = \alpha_{y_1} + \beta_{11}L_1 + \beta_{12}L_2 + \beta_{13}L_1L_2 + \zeta_1$. If we rewrite this in observed variables, we get Equation 10, $y_1 = \alpha_{y_1} + \beta_{11}x_1 + \beta_{12}x_3 + \beta_{13}x_1x_3 + u_1$, and the IVs are x_2, x_4, and x_2x_4. Next, we consider a number of extensions to this basic model and how the IVs would change in each case.

Extension 1: More Indicators for Explanatory Latent Variables. Keeping the same equation with latent variables, how do things change when we have more than two indicators for one or more of the latent variables? As before, we construct product variables by multiplying each indicator of the first latent variable by each indicator of the second latent variable. Figure 6.2a has an example to illustrate this. The latent variable model is the same as in Fig. 6.1. But we have three indicators of the first latent variable and four indicators of the second. Those products that involve the indicator that scales each of the latent variables will not be eligible to be IVs because they will correlate with the composite disturbance term, u_1, in Equation 10. The

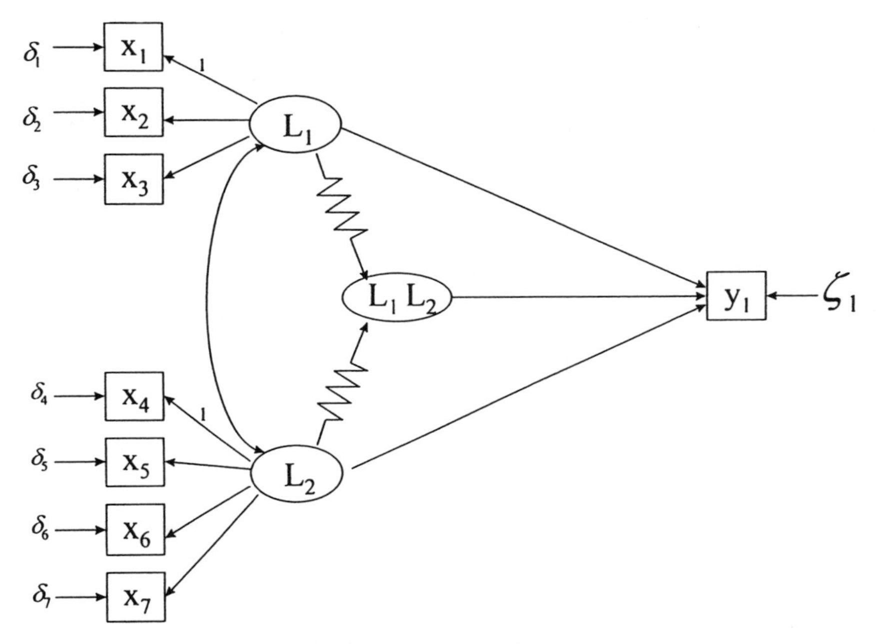

FIG. 6.2a. More indicators for independent latent variables.

product variables that are available as IVs are: x_2x_5, x_2x_6, x_2x_7, x_3x_5, x_3x_6, and x_3x_7. Add to this the linear nonscaling indicators, x_2, x_3, x_5, x_6, and x_7, and we have a total of 11 IVs. In Equation 10 we have only three right-hand-side variables to replace (x_1, x_3, and x_1x_3). In this situation, we have more than enough IVs and the equation is overidentified. The exact same 2SLS procedure is followed as before except the list of IVs has expanded.

Extension 2: More Indicators for Dependent Latent Variables. Suppose that unlike Equation 1, the dependent variable is latent and measured with multiple indicators. Figure 6.2b shows this change in the model where we include three indicators of L_4, the latent dependent variable. In this case the list of IVs remains unchanged. The reason is that ζ_4 is part of the composite disturbance term, u_1, and because ζ_4 has a direct effect on L_4, all of the indicators of L_4 are correlated with ζ_4 and thus with u_1. None of these new indicators is suitable. So in the model in Fig. 6.2b, the list of IVs is the same as in Fig. 6.1.

Extension 3: A Second Dependent Latent Variable. Another possibility is that the dependent variable in the equation with the product interaction of latent variables might influence other latent or observed variables in the system. Consider Fig. 6.2c to illustrate this point. We have modified Fig. 6.2b to allow L_4 to influence another latent variable, L_5, which in turn has

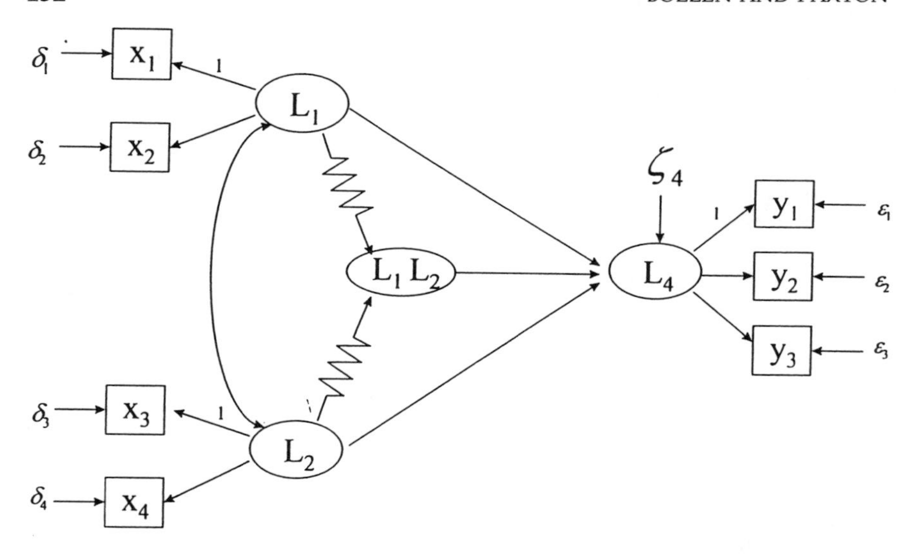

FIG. 6.2b. More indicators for dependent latent variable.

two indicators. None of the observed variables that is directly or indirectly influenced by L_4 is eligible as an IV in the L_4 equation. The reason is that the disturbance term for L_4 has an indirect effect on all of the variables that L_4 affects and since the disturbance, ζ_4, appears in the composite disturbance, u_1, this disqualifies all variables that L_4 influences. Another way to determine which variables are disqualified is to trace the arrows in the path diagram.

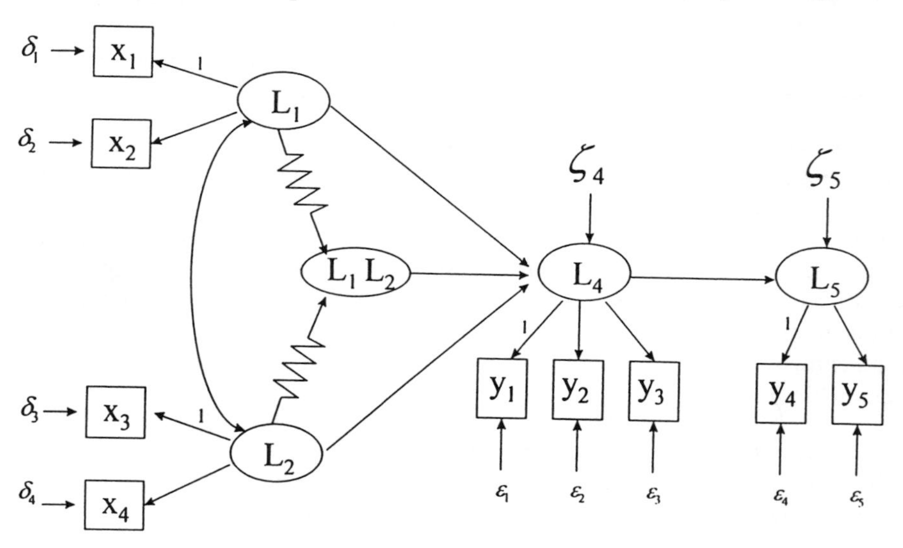

FIG. 6.2c. A second dependent latent variable.

If the researcher can trace a path following arrows from a linear term in the composite disturbance to a variable, then that variable is disqualified. Thus, one could follow arrows from ζ_4 to y_4 and y_5 so they are both disqualified.

Extension 4: Additional Exogenous Variables. Typically a model with interactions of the latent variables includes other exogenous variables as well. Such additional exogenous variables are a source of additional IVs. For instance, suppose that we add two exogenous observed variables to Equation 1,

$$y_1 = \alpha_{y_1} + \beta_{11}L_1 + \beta_{12}L_2 + \beta_{13}L_1L_2 + \gamma_{11}z_1 + \gamma_{12}z_2 + \zeta_1 \tag{12}$$

where z_1 and z_2 are exogenous variables that are independent of all disturbances and error terms. Even after the scaling indicators and their errors of measurement are substituted in to replace the latent variables, z_1 and z_2 remain uncorrelated with the composite disturbance. They can serve as IVs. This generalizes to any number of exogenous observed variables: They can serve as IVs along with the other IVs for an equation.

The situation is similar for the nonscaling indicators of latent exogenous variables that are not part of the product interaction and for which the errors of the indicators are uncorrelated with the errors of the indicators of the latent variables that are part of the product interaction. Figure 6.3 gives an example. It is the same as Fig. 6.1 except we have added a latent exogenous variable, L_3, that has two indicators, x_5 and x_6. The x_5 scales L_3 and x_5's error will be part of the new composite disturbance. This makes x_5 ineligible as an IV. But x_6 is eligible. Furthermore, if there were additional indicators of L_3, they too could serve as IVs. If we added another latent exogenous variable, all but the indicator that scales this new latent variable would be suitable IVs.[3]

It might occur to the reader that we could create even more IVs by multiplying or forming other functions of all the variables that qualify as IVs when taken alone. We generally do not recommend this step if a sufficient number of IVs are available without doing this. The only exception to this advice is for the indicators of the latent variables that are part of the interaction term. Researchers should use the product of all possible pairs of nonscaling indicators where the first indicator of the pair is from the first latent variable and the second is from the second latent variable. Any pair that includes the scaling indicators of these latent variables should be excluded. These products of indicators provide valuable information because they are reflecting the product interaction of the latent variables.

If we restrict ourselves to models with product interactions of latent variables and ones where errors of measurement are uncorrelated, the pre-

[3]We again assume that the errors for these new indicators are uncorrelated with the errors of the indicators of the latent variables that are part of the interaction term.

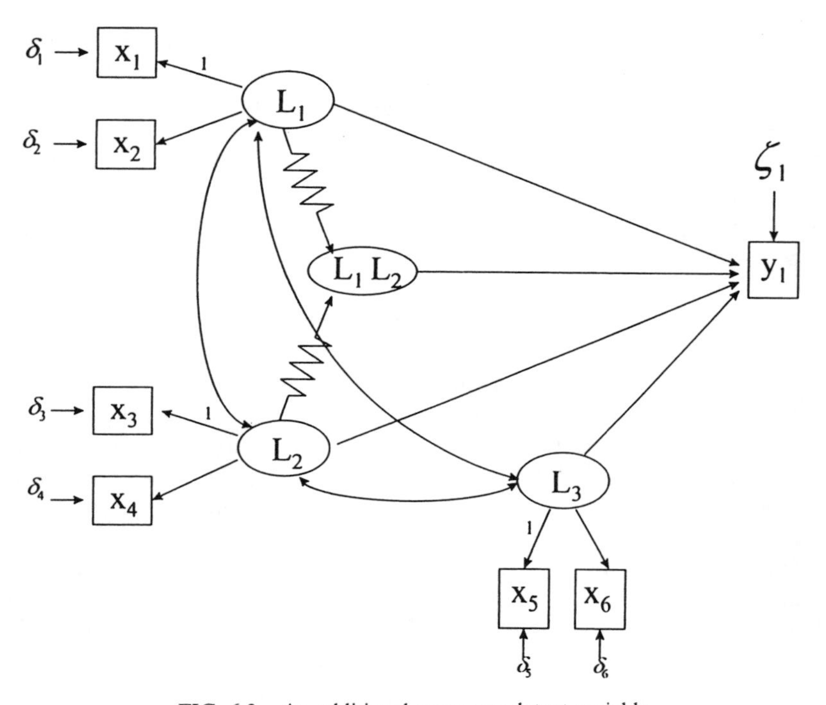

FIG. 6.3. An additional exogenous latent variable.

ceding discussion provides a procedure for selecting IVs in the type of models encountered most often in practice. Table 6.1 summarizes these rules for including and excluding observed variables as IVs in the equation with the product latent variable term.

These rules are not comprehensive. For instance, if there are correlated errors of measurement between the errors of the indicators that are part of the latent variables that are in the product interaction, then some more variables might be excluded due to this correlated error. Other complications could arise if a model has variables with a factor complexity greater than one or if the observed variables affect one another. A more technical discussion of the conditions for IVs in models with nonlinear functions of latent variables is in Bollen (1995), but the most typical cases encountered in practice are covered by the rules we describe here.

Evaluating Instrumental Variables

We briefly mention a few other aspects of selecting IVs in these models. One is a simple counting rule that is tied to the identification of the equation. As

TABLE 6.1
Rules for Selecting Instrumental Variables for an Equation With a Product Interaction of Latent Variables

Rules for Including IVs
1. Take each nonscaling indicator of the first latent variable and multiply it by one of the nonscaling indicators of the second latent variable in the product interaction. Repeat this step until all possible pairs of indicators are formed and use these as IVs.
2. Include all exogenous observed variables as IVs.
3. Include all nonscaling indicators of exogenous latent variables that are on the right-hand side of the equation but are not part of the product interaction term.
4. Use as IVs, all indicators of exogenous latent variables that are not in the equation.
Rules for Excluding IVs
1. Do not select as IVs any variables that scale the latent variables that are in the equation to be estimated.
2. Exclude all observed variables that are directly or indirectly affected by the dependent variable of the equation with the product interaction in it:
a. If the dependent variable is latent, then do not use its indicators.
b. Any other observed variables that are directly or indirectly functions of this dependent variable are not eligible.

suggested earlier, we must have at least as many IVs as we have variables on the right-hand side of the equation that we will estimate. In the earlier example of Equation 9, we had three right-hand-side variables so we needed at least three IVs. If we have eight variables on the right-hand side, then we need at least eight IVs. Too few IVs suggests that the equation is underidentified for the 2SLS method.

Even if we satisfy the counting rule, the IVs might fall short of the ideal. Generally, we want to select IVs that do a good job of predicting the variables that they will replace. The R^2's from the regression of each right-hand-side variable on the full set of IVs are one gauge of quality. An R^2 that is low (e.g., less than .10) could affect the quality of the estimates. We recommend that researchers routinely calculate and examine the R^2's from the first stage of the 2SLS estimates.

Another way to assess the appropriateness of the IVs is applicable when there are more IVs than right-hand-side variables, that is, when the equation is overidentified. Several tests of overidentification are available and these provide evidence on whether the IVs are really uncorrelated with the disturbance of the equation. For instance, Proc Syslin in SAS provides the Basmann (1960) test of overidentification as an option. We recommend that one or more of these tests be routinely employed in cases where the equation is overidentified.

THE KENNY AND JUDD METHOD

Another type of multiple-indicator technique began with the work of Kenny and Judd (1984). Like the 2SLS approach, their method involves the formation of cross-products of the indicators of the two latent variables involved in the interaction. (In the earliest example, Equations 1 to 5, the cross-products are x_1x_3, x_1x_4, x_2x_3, and x_2x_4.) Kenny and Judd used the information gained from the cross-products in a different way than the 2SLS approach, however. Kenny and Judd included the products as indicators of the interaction latent variable in a series of additional equations in the measurement model. In our running example, four new measurement equations would be formed:

$$x_1x_3 = L_1L_2 + L_1\delta_3 + L_2\delta_1 + \delta_1\delta_3 \quad (13)$$

$$x_1x_4 = \lambda_{41}L_1L_2 + L_1\delta_4 + \lambda_{41}L_2\delta_1 + \delta_1\delta_4 \quad (14)$$

$$x_2x_3 = \lambda_{21}L_1L_2 + L_2\delta_2 + \lambda_{21}L_1\delta_3 + \delta_2\delta_3 \quad (15)$$

$$x_2x_4 = \lambda_{21}\lambda_{41}L_1L_2 + \lambda_{21}L_1\delta_4 + \lambda_{41}L_2\delta_2 + \delta_2\delta_4. \quad (16)$$

These new equations require the imposition of linear and nonlinear constraints on the estimates. For example, in Equation 14 the factor loadings for L_1L_2 and that for L_1 in the x_4 equation (5) must be equal, because they are both equal to λ_{41}. Equation 16 has a nonlinear constraint on the coefficient for the L_1L_2 variable.

Unlike the 2SLS method, the Kenny and Judd (1984) method and its variants assume that the observed noninteraction variables come from normal distributions. Under this assumption, additional constraints are required in the covariance matrix of the latent variables, such as, $\text{VAR}(L_1L_2) = \text{VAR}(L_1)\text{VAR}(L_2) + [\text{COV}(L_1,L_2)]^2$. Constraints in other matrices, such as the covariance of errors, are needed as well (see Jaccard & Wan, 1996, for a complete listing of constraints). If the new measurement equations and nonlinear constraints are included, Kenny and Judd showed that the coefficients for the nonlinear variables could be consistently estimated with generalized least squares (GLS) estimation.

There are some problems with the Kenny and Judd (1984) technique. First, the introduction of the product indicators introduces non-normality into the system of equations, so the standard errors are in question. Second, their technique requires that the variables be in deviation form, so intercepts are not estimated. Additionally, the technique introduces many new product terms and corresponding nonlinear constraints.

Due to the nonlinear constraints, the technique was originally extremely hard to implement within existing software packages. This problem was addressed by Hayduk (1987), who introduced a series of latent variables into

LISREL to "trick" it into estimating the coefficients. Such a trick was no longer necessary when PROC CALIS in SAS and LISREL 8 were introduced because they allowed for both linear and nonlinear constraints. As Jöreskog and Yang (1996) pointed out, however, even though it is now possible to program the constraints, it is still tricky but absolutely essential to get them right. If the constraints are calculated or programmed incorrectly, severe consequences can result because the model will not be correctly specified. Furthermore, the technique requires variables that are from normal distributions, a condition rarely obtained in practice.

Variant 1: Fewer Product Terms. To deal with some of these problems, two variants of the Kenny and Judd (1984) method arose. The first variant involved restricting the number of product indicators used in the model, which in turn reduces the number of constraints and the number of non-normally distributed product variables. Jaccard and Wan (1996) first used this technique in their simulated comparison of the maximum likelihood estimator (MLE) and the asymptotic distribution function (ADF; also called WLS). Thus, although there were nine cross-product indicators available for their interaction latent variable, they used only four.

Jöreskog and Yang (1996) took the reduction of the number of product indicators further and proved that the model is identified with only a single product indicator. Under their method, only one new equation is added to the measurement model. The product indicator they used is the product of the two scaling indicators:

$$x_1x_3 = L_1L_2 + L_1\delta_3 + L_2\delta_1 + \delta_1\delta_3. \quad (17)$$

In addition to reducing the number of constraints required, they noted that restricting to a single product indicator reduced the number of variables from non-normal distributions used in the system. They also emphasized the importance of including information on the means of the observed variables in the estimation, arguing that making use of the means and intercepts is actually essential because the means are a function of the other variables in the model so to set them to zero is an arbitrary choice. So, their technique also allows the estimation of intercepts. Jöreskog and Yang presented a variant of a WLS estimator to produce standard errors that are asymptotically correct. They also noted, however, that using that estimator can lead to convergence problems.

Variant 2: A Two-Step Procedure. The second variant to the Kenny and Judd (1984) technique was proposed by Ping (1994, 1996). His approach retains all of the possible product terms and uses a two-step technique in estimation. Briefly, the technique involves: (a) estimating the model without the interaction term, (b) algebraically calculating the factor loadings and error variances for the products of the nonlinear terms, (c) fixing the loadings at the calculated values while estimating the other coefficients, and (d) repeating Steps 2 and 3 until there is little change in the parameters.

This variant also exhibits a number of problems. The errors must be uncorrelated with each other and it requires normality of the observed variables. Additionally, the approach does not correct the standard errors for the constrained parameters.

COMPARING THE METHODS

Although the 2SLS and the Kenny and Judd (1984) methods both make use of product indicators, they do so in very different ways. Thus, it is useful to compare the two methods on a number of criteria. Table 6.2 presents some criteria that are related to the usefulness of the techniques. For example, nonlinear terms besides squares and interactions are difficult to implement in the Kenny and Judd–type methods but fairly easy to implement in the 2SLS procedure.[4] The 2SLS procedure can also handle non-normal observed variables whereas the Kenny and Judd and related techniques cannot. Additionally, as noted earlier, the inclusion of product terms as indicators actually adds non-normality into the system for the product indicator models. The asymptotic distribution of the 2SLS estimator is known; Kenny and Judd did not provide the asymptotic distribution of their estimator. But, Jöreskog and Yang (1996) described asymptotic properties for their modified Kenny and Judd procedure when the nonproduct variables are from normal distributions and the means are part of the analysis. The small to moderate sample properties are unknown for all the alternatives. Additional differences between the techniques are noted in the table. For example, with the 2SLS method it is not necessary to include nonlinear constraints, construct a measurement model for the product indicators, or use a special software package. On the other hand, for the Kenny and Judd technique, it is not necessary to use instrumental variables.

Comparison With Kenny and Judd's Simulated Data

To illustrate their method, Kenny and Judd (1984) simulated a model with an interaction of latent variables. Their model is exactly the same as our example shown in Fig. 6.1. They simulated 500 cases with the following population parameters:

[4]Higher-order polynomials, such as cubes, require some special considerations (see Bollen, 1995).

TABLE 6.2
Comparison of the Two Methods

	Kenny and Judd–Type Methods			*2SLS*
	(Kenny and Judd)	*(Ping)*	*(Jöreskog and Yang)*	*(Bollen)*
Allows multiple indicators?	yes	yes	yes	yes
Allows unknown reliabilities?	yes	yes	yes	yes
Designed for nonlinear functions of latent variables:				
in latent variable model?	yes	yes	yes	yes
in measurement model?	no	no	no	yes
Applicable to non-linear terms besides squares and interactions?	no	no	no	yes
Can handle non-normal observed variables?	no	no	no	yes
Asymptotic distribution of estimator known?	no	no	yes	yes
Small sample properties known?	no	no	no	no
No need to include nonlinear constraints?	no	no	reduced need	yes
Estimates intercepts?	no	no	yes	yes
Can estimate only the equation with the nonlinear term?	no	no	no	yes
No need to construct measurement model for product indicators?	no	no	no	yes
Can estimate with standard procedures in SAS, SPSS, STATA?	Proc Calis	Proc Calis	Proc Calis	all packages
Assumes independence of latent variables and errors?	yes	yes	yes	yes
No need to form instrumental variables?	yes	yes	yes	no

$$y_1 = -.15L_1 + .35L_2 + .70L_1L_2 + \zeta_1 \quad (18)$$

$$x_1 = L_1 + \delta_1 \quad (19)$$

$$x_2 = .60L_1 + \delta_2 \quad (20)$$

$$x_3 = L_2 + \delta_3 \quad (21)$$

$$x_4 = .70L_2 + \delta_4. \quad (22)$$

We refer the reader to Equations 1 through 11 to understand the substitution process for this model. To estimate the parameters, they use GLS where the weighting matrix is the inverse of the sample covariance matrix. They also presented the covariance matrix for their simulated data. Using that matrix, we can get the 2SLS estimates of the parameters. The first-stage R-squares for the three endogenous variables are .43, .46, and .16. None of these is extremely low (below .10) which indicates that the instrumental variables are of decent "quality." The SAS program for the simulated data example is in the Appendix.

In Table 6.3 we present a comparison of the two sets of estimates for the latent variable equation. For both methods, the estimated coefficients are close to the population coefficients. The same result is obtained for B_{13}, whereas for B_{11} and B_{12}, the 2SLS procedure produces an estimate closer to the original than the Kenny and Judd (1984) procedure. Additionally, the 2SLS estimator provides standard errors for the estimates and these indicate that all of the 2SLS estimates are significant. Kenny and Judd did not report standard errors.

Comparison With Kenny and Judd's Empirical Data

Kenny and Judd (1984) also provided a complex empirical example of voters' agreement with candidates, based on whether they like or dislike the candidates. Kenny and Judd explained that the process of assimilation would

TABLE 6.3
Comparison of the Simulation Results

Coefficient	*Generated Value*	*Kenny and Judd Estimate*	*2SLS Estimate*
gamma11	−0.150	−0.169	−0.160 (0.052)
gamma12	0.350	0.321	0.360 (0.054)
gamma13	0.700	0.710	0.710 (0.053)

Note. Standard errors are in parentheses.

involve a voter overestimating their agreement with a candidate, if they like the candidate. Conversely, contrast occurs when voters underestimate their agreement (or overestimate their disagreement) with a candidate whom they do not like. Thus, a voter's judgment of the candidate's position (C) is based on his or her own position on an issue (V) and should be moderated by his or her feelings toward the candidate (S). This means we would expect an interaction of V and S—VS, in the equation for candidate's position (C).

Additionally, Kenny and Judd (1984) noted that research indicates that assimilation is more powerful than contrast. Thus, we would also expect an effect of the square of the interaction term, VS^2, so that the effect of the interaction is stronger for higher levels. From this theoretical base, Kenny and Judd estimated the following model,

$$C = \beta_{11}V + \beta_{12}S + \beta_{13}S^2 + \beta_{14}VS + \beta_{15}VS^2 + \zeta_1 \tag{23}$$

with data taken from the 1968 National Election Survey. The closeness of voters to two candidates, Richard Nixon and Hubert Humphrey, on two issues, the Vietnam War and control of crime, are measured. Thus, there are two indicators for C, c_1 and c_2, indicating the voters' judgments of the candidates on the Vietnam War and crime, respectively. V has two indicators, v_1 and v_2, indicating the voters' position on those two issues as well. They treated sentiment, S, as a perfectly measured variable on a 1 to 100 (feeling thermometer) scale. The sample size was 1,160. For further details on the measurement of the variables, we direct the reader to Kenny and Judd (1984).

As with the simulated data, Kenny and Judd (1984) presented the covariance matrix for the empirical example in their article. Thus, we reestimated the model using the 2SLS technique. The SAS program using Proc Syslin is listed in the Appendix. The results of the two estimation procedures are compared in Table 6.4. From the table, one can see that the two sets of estimates are similar, though further apart than the simulated data. The first-stage R-squares for the 2SLS technique are a bit low at .11, .15, and .23, but not excessively so.

Although Kenny and Judd (1984) did not present standard errors for their estimates, the 2SLS procedure does provide them, so we can say something about the statistical significance of the estimates.[5] The estimate for the interaction term is significant and so the evidence supports the contention that a voter's perceived agreement with a candidate is moderated by his or her feelings about the candidate. The VS^2 is not significant, so we do not have evidence for this effect being greater at higher levels.

[5]Because Kenny and Judd (1984) used mean deviated scores and did not report means, we cannot model the intercepts. We must be cautious in our interpretations of significance, because we are not accounting for that information.

TABLE 6.4
Comparison of the Empirical Results

Coefficient	*Kenny and Judd Estimate*	*2SLS Estimate*
β_{11}	0.180	0.151 (.100)
β_{12}	−0.111	−0.191 (.025)
β_{13}	−0.019	−0.032 (.007)
β_{14}	0.207	0.255 (.068)
β_{15}	0.009	0.014 (.012)

Note. $N = 1{,}160$. Standard errors are in parentheses.

AN EMPIRICAL EXAMPLE

To illustrate the use of the 2SLS method for more complicated models, we provide an empirical example that utilizes a larger model. Suppose we are interested in the societal condition of normlessness, anomie (Durkheim, 1897/1951), and its individual experience, anomia (Srole, 1956), which is distinguished by a loss of orientation and feelings of apathy. There are a number of demographic variables that could be associated with higher anomia, such as gender, race, education, income, and religion. Additionally, we might be interested in the potential impact of a respondent's overall trust in individuals, their confidence in various institutions, and the interaction of the two on anomia. The path diagram of such a model is shown in Fig. 6.4. The data were taken from the 1976 General Social Survey (Davis & Smith, 1994).

Variables Included in the Model

The dependent variable (y_1) is a scale created from five anomia questions:

1. You sometimes can't help wondering whether anything is worthwhile any more.
2. In spite of what some people say, the lot (situation/condition) of the average man is getting worse, not better.
3. It's hardly fair to bring a child into the world with the way things look for the future.
4. Most public officials (people in public office) are not really interested in the problems of the average man.
5. Most people don't really care what happens to the next fellow.

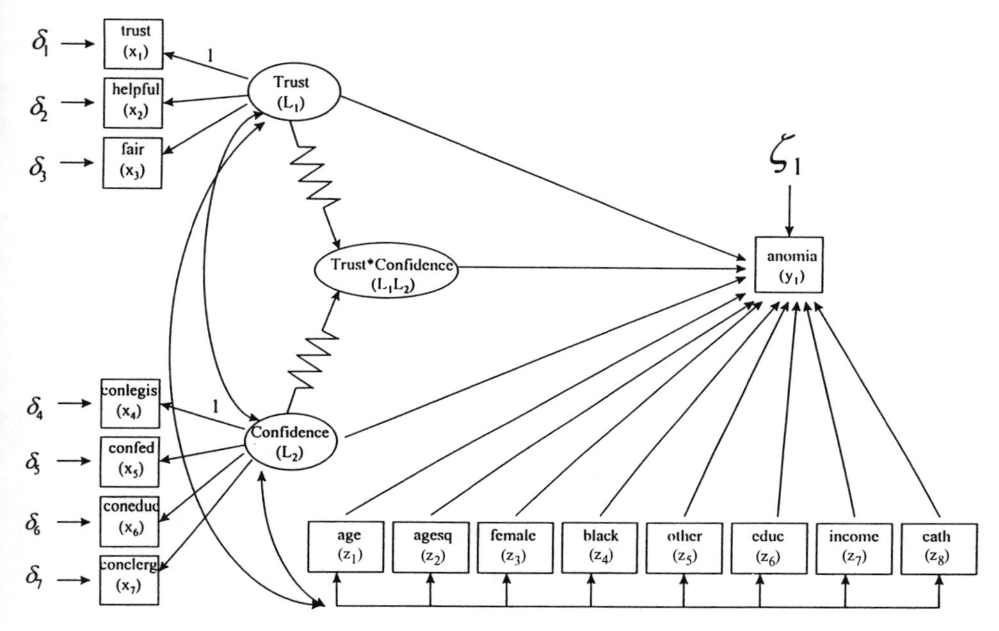

FIG. 6.4. An empirical example.

Each of the questions is coded as a dichotomy, with a "yes" answer indicating greater anomia. To create the variable y_1, all yes answers were added together and the total divided by five, resulting in a variable that ranges from zero to one, with higher numbers indicating greater anomia.

The first latent variable (L_1), trust in individuals, is measured with three indicators:

1. Generally speaking, would you say that most people can be trusted or that you can't be too careful in dealing with people (x_1)?
2. Would you say that most of the time people try to be helpful, or that they are mostly just looking out for themselves (x_2)?
3. Do you think most people would try to take advantage of you if they got a chance, or would they try to be fair (x_3)?

Each indicator has three categories but the middle category, "depends," contains very few cases. The second latent variable (L_2), confidence in institutions, has four indicators, each with three categories, measuring confidence in Congress (x_4), the executive branch of the federal government (x_5), education (x_6), and organized religion (x_7).

Eight exogenous variables are also included in the model: age (z_1), age squared (z_2), female (z_3), race (indicated by two dummy variables, Black [z_4]

and other [z_5]), education (z_6), income (z_7), and Catholic (z_8). Catholic is included to represent the distinction between Protestantism and Catholicism in Durkheim's classic study of suicide. The square of age is included in the model to capture a potential nonlinear relationship with anomia (y_1).

The equation with the interaction of latent variables for this empirical example is a simple extension of Equation 12:

$$y_1 = \alpha_{y_1} + \beta_{11}L_1 + \beta_{12}L_2 + \beta_{13}L_1L_2 + \gamma_{11}z_1 + \gamma_{12}z_2$$
$$\ldots + \gamma_{17}z_7 + \gamma_{18}z_8 + \zeta_1. \tag{24}$$

The measurement equations are:

$$x_1 = L_1 + \delta_1 \tag{25}$$
$$x_2 = \alpha_2 + \lambda_{21}L_1 + \delta_2 \tag{26}$$
$$x_3 = \alpha_3 + \lambda_{31}L_1 + \delta_3 \tag{27}$$
$$x_4 = L_2 + \delta_4 \tag{28}$$
$$x_5 = \alpha_5 + \lambda_{52}L_2 + \delta_5 \tag{29}$$
$$x_6 = \alpha_6 + \lambda_{62}L_2 + \delta_6 \tag{30}$$
$$x_7 = \alpha_7 + \lambda_{72}L_2 + \delta_7 \tag{31}$$

where the first three equations are for the indicators of trust (L_1) and the last four are for the measures of confidence in institutions (L_2).[6] We chose as scaling indicators those variables that seemed theoretically the closest to the latent variables. Our theoretical decision is supported by empirical evidence as well. We regressed each possible scaling variable ($x_1 - x_7$) on all the other variables in the model and the chosen scaling variables had the highest R-squares, though the differences in R-squares were not always great.

Estimating the Model With 2SLS

To estimate the model using 2SLS, we first substitute each scaling indicator for its latent variable in the latent variable equation. This produces the following equation to be estimated with the 2SLS procedure.

---

[6]We ignore the ordinal nature of the indicators and treat them as if they were continuous for the sake of this illustration.

$$y_1 = \alpha_{y_1} + \beta_{11}x_1 + \beta_{12}x_4 + \beta_{13}x_1x_4 + \gamma_{11}z_1 + \gamma_{12}z_2$$
$$\ldots + \gamma_{17}z_7 + \gamma_{18}z_8 + u_1. \quad (32)$$

The disturbance, u_1, is a composite disturbance equal to:

$$u_1 = -\beta_{11}\delta_1 - \beta_{12}\delta_4 - \beta_{13}(\delta_1x_4 - \delta_4x_1 + \delta_1\delta_4) + \zeta_1. \quad (33)$$

The exogenous perfectly measured variables (the *z*s) represent themselves as IVs and are not associated with the disturbance term. The x_1, x_4, and x_1x_4 variables are associated with the disturbance term and will need to be predicted with IVs.

We include all of the exogenous variables and each of the nonscaling indicators of the two latent explanatory variables as IVs. To obtain more IVs, we create the products of the nonscaling indicators associated with the two latent variables in the interaction. Multiplying across the latent variables means that we will have 2 × 3 terms to include as IVs. In this example, these are x_2x_5, x_2x_6, x_2x_7, x_3x_5, x_3x_6, and x_3x_7. The SAS code for this example appears in the appendix.

The Results

The results appear in Table 6.5. Overall, the high first-stage *R*-square values (.32, .21, and .27) and the nonsignificant test for overidentification support our choice of IVs. The coefficient for the interaction of trust in individuals and confidence in institutions indicates that it is a statistically significant predictor of an individual's feelings of anomia with a one-tailed test.[7] This in turn means that the coefficients of the two variables, trust and conlegis, cannot be interpreted independently. The negative impact of either one on anomia is conditional on the level of the other. The interaction term indicates the change in the slope of the relation between one component of the interaction and anomia for a one-unit shift in the other component. We represent the changes in slope, and the consequent change in anomia in a table. Table 6.6 shows the expected change in anomia for a one-unit change in one part of the interaction, with the other part of the interaction at its minimum, at .5, and its maximum. The main effects of the two components of the interaction, trust and conlegis, can only be interpreted individually if one is held at zero. Thus, if an individual's confidence in the legislature is at zero, then a one-unit change in trust is expected to decrease anomia by .324.

[7]If three outlying cases are removed (id = 80, 542, 647) then the interaction becomes significant under a two-tailed test ($p = .04$).

TABLE 6.5
2SLS Results for Anomia Empirical Example

Variable	*Parameter Estimate (Standard Error)*
Intercept	0.710***
	(.111)
Trust (x_1)	–.324**
	(.116)
Conlegis (x_4)	–.067
	(.146)
Trust*conlegis (x_1x_4)	–.484†
	(.259)
Age (z_1)	.010***
	(.003)
Agesq (z_2)	–.0001**
	(.0000)
Female (z_3)	.024
	(.019)
Black (z_4)	.006
	(.037)
Other (z_5)	–.051
	(.118)
Educ (z_6)	–.008†
	(.004)
Income (z_7)	–.005***
	(.001)
Cath (z_8)	0.015
	(.021)

Note. $N = 1{,}137$. Overidentification test: F value = 1.32, $df1 = 8$, $df2 = 1{,}117$; p value = .23. †$p < .10$. * = $p < .05$. ** = $p < .01$. *** = $p < .001$.

As for the demographic variables, age and age^2 are both significant, and support the contention that there is a nonlinear, inverted U effect of age on anomia. Income is significantly negative, indicating that as an individual's income increases, their feeling of anomia decreases. Education is significant with a one-tailed test, with higher levels of education reducing the expected value of an individual's feelings of anomia.

Substantively, the interaction has a reasonably strong impact on anomia: If trust and confidence in the legislature move from their lowest level to their highest level, there is an expected decrease in anomia of .804 on a scale that ranges from 0 to 1. There are other indications that the interaction does not make an important contribution to the model of anomia, however. First, comparing the adjusted R-squares for the model with the interaction term (.28) and without the interaction term (.29) indicates that the interaction term does not add any explanatory potential to the model. Second, the same model, replicated for 1973 data, does not produce a statistically significant interaction term, although the main effects of the variables in the interaction

TABLE 6.6
Interpreting the Interaction Term

Change in Confidence (x_4)	*Value of Trust* (x_1)	*Expected Change in Anomia* (y_1)
1	0	−.07 + −.484(0) = −.07
1	.5	−.07 + −.484(.5) = −.312
1	1	−.07 + −.484(1) = −.554
Change in Trust (x_1)	*Value of Confidence* (x_4)	*Expected Change in Anomia* (y_1)
1	0	−.32 + −.484(0) = −.32
1	.5	−.32 + −.484(.5) = −.562
1	1	−.32 + −.484(1) = −.804

are significant. The test for overidentification for that year is .06, and for 1976 without an interaction it is .07, indicating that there may be a problem with the IVs.

Handling Correlated Errors

Thus far, we have assumed that all of the indicators of the latent variables have uncorrelated disturbance terms. However, if a researcher believes that two of the indicators of the latent variable are correlated, this can also be handled with the 2SLS technique. If the correlation is between two of the disturbances of nonscaling indicators for a single latent variable (e.g., coneduc and conclerg), then there is no impact on the model and it would be estimated as earlier. However, if the correlation is between the errors for a scaling indicator and a nonscaling indicator (e.g., conlegis and confed), then the variable correlated with the scaling indicator, in this case confed, becomes ineligible as an IV. Other possible combinations of correlation can occur. But the key idea is to see whether the correlated error leads to a correlation between the composite disturbance and the candidate IV.

CONCLUSIONS

Interactions are part of many substantive areas of research. It is essential that SEMs have suitable methods for incorporating them into models. In this chapter we have focused on a 2SLS method to handle interactions of latent variables that have multiple indicators. The availability of 2SLS routines in the most common statistical packages (e.g., SPSS, SAS, STATA) means that specialized software is not required. The method does require the selection of IVs, and we have given rules for their selection in the most common cases. Significance testing with asymptotic standard errors is possible because the 2SLS estimator has a known asymptotic distribution and it does not require that the observed variables come from multinormal distributions. In addition, there are diagnostic statistics to help in assessing the quality of the IVs.

Along with the merits of the method, we wish to point out its limitations. First, as with the other estimators in SEM, the 2SLS estimator's properties are largely justified through asymptotic theory. We need evidence to assess its performance in finite samples. The literature on 2SLS from the econometric literature on simultaneous equation models with observed variables is largely favorable (see e.g., Johnston, 1972; Kennedy, 1985), but we do not have as extensive a literature on it in latent variable models. Second, the selection of the IVs for the model depends on the correctness of the structural equation model. That is, the selection of the IVs depends on the specification of the model and what the specification implies about the correlation of observed variables with the disturbance in the equation that we are to estimate. Changes in the model can change the status of a variable as a suitable IV. In equations that are overidentified, we have test statistics to evaluate the appropriateness of IVs, but in exactly identified models this will not be true.[8]

Table 6.2 summarizes the major features of each technique. A more complete evaluation of each awaits further analytical and Monte Carlo simulation research.

ACKNOWLEDGMENTS

Presented at the Social Science and Statistics Conference in Honor of Clifford C. Clogg at Pennsylvania State University, State College, Pennsylvania, September 26–28, 1996. This paper was published in *Structural Equation Modeling: A Multidisciplinary Journal*, 5(3), 267–293, 1998.

REFERENCES

Basmann, R. L. (1960). On finite sample distributions of generalized classical linear identifiability test statistics. *Econometrica, 45,* 939–952.

[8]Full-information methods such as the Kenny and Judd (1984) method also will be affected by misspecified relations in a SEM. They are likely to be even more sensitive to model specifications because they estimate all parameters simultaneously. Full-information estimators can spread specification error in one part of the system to other parts. The 2SLS estimator appears to better isolate specification error to the equations in which it occurs (Cragg, 1967).

Bollen, K. A. (1995). Structural equation models that are nonlinear in latent variables: A least squares estimator. In P. M. Marsden (Ed.), *Sociological methodology, 1995* (pp. 223–251). Cambridge, MA: Blackwell.

Bollen, K. A. (1996). An alternative two stage least squares (2SLS) estimator for latent variable equations. *Psychometrika, 61,* 109–121.

Busemeyer, J. R., & Jones, L. E. (1983). Analysis of multiplicative combination rules when the causal variables are measured with error. *Psychological Bulletin, 93,* 549–562.

Cragg, J. (1967). On the relative small-sample properties of several structural equation estimators. *Econometrica, 35,* 89–110.

Davis, J. A., & Smith, T. W. (1994). *General social surveys, 1972–1994* [Machine-readable data file]. Chicago: National Opinion Research Center [Producer]. Storrs, CT: The Roper Center for Public Opinion Research, University of Connecticut [Distributor]. [One data file (32,380 logical records) and one codebook (1,073 pp.)]

Durkheim, E. (1951). *Suicide: A study in sociology.* New York: The Free Press. (Original work published 1897)

Feucht, T. E. (1989). Estimating multiplicative regression terms in the presence of measurement error. *Sociological Methods and Research, 17,* 257–282.

Hayduk, L. A. (1987). *Structural equation modeling with LISREL.* Baltimore: Johns Hopkins University Press.

Heise, D. R. (1986). Estimating nonlinear models correcting for measurement error. *Sociological Methods and Research, 14,* 447–472.

Jaccard, J., Turrisi, R., & Wan, C. H. (1990). *Interaction effects in multiple regression* (Sage University Paper Series on Quantitative Applications in the Social Sciences, No. 07-072). Thousand Oaks, CA: Sage.

Jaccard, J., & Wan, C. H. (1996). *LISREL approaches to interaction effects in multiple regression* (Sage University Paper Series on Quantitative Applications in the Social Sciences, No. 07-114). Thousand Oaks, CA: Sage.

Johnston, J. (1972). *Econometric methods* (2nd ed.). New York: McGraw-Hill.

Jöreskog, K. G., & Yang, F. (1996). Nonlinear structural equation models: The Kenny–Judd model with interaction effects. In G. A. Marcoulides & R. E. Schumacker (Eds.), *Advanced structural equation modeling* (pp. 57–88). Mahwah, NJ: Lawrence Erlbaum Associates.

Kennedy, P. (1985). *A guide to econometrics.* Cambridge, MA: MIT Press.

Kenny, D. A., & Judd, C. M. (1984). Estimating the non-linear and interactive effects of latent variables. *Psychological Bulletin, 96,* 201–210.

Ping, R. A. (1994). Does satisfaction moderate the association between alternative attractiveness and exit intention in a marketing channel? *Journal of the Academy of Marketing Science, 22,* 364–371.

Ping, R. A. (1996). Latent variable interaction and quadradic effect estimation: A two-step technique using structural equation analysis. *Psychological Bulletin, 119,* 166–175.

SAS Institute Inc. (1988). *SAS/ETS user's guide.* Cary, NC: Author.

Srole, L. (1956). Social integration and certain corollaries: An exploratory study. *American Sociological Review, 21,* 709–716.

APPENDIX

SAS Programs (see SAS Institute, 1988, for Proc Syslin)

```
SAS PROGRAM FOR KENNY AND JUDD SIMULATED DATA (N=500)
data a (type=cov);
infile cards missover;
```

```
input _name_ $ _type_ $ X1 X2 X3 X4 X1X3 X1X4 X2X3 X2X4 Y1;
cards;
X1   COV 2.395
X2   COV 1.254 1.542
X3   COV  .445  .202 2.097
X4   COV  .231  .116 1.141 1.370
X1X3 COV -.367 -.070 -.148 -.133 5.669
X1X4 COV -.301 -.041 -.130 -.117 2.868 3.076
X2X3 COV -.081 -.054  .038  .037 2.989 1.346 3.411
X2X4 COV -.047 -.045  .039 -.043 1.341 1.392 1.719 1.960
Y1   COV -.368 -.179  .402  .282 2.556 1.579 1.623  .971 2.174
N    N    500   500   500   500   500   500   500   500   500
;
run;
proc print;
proc syslin 2sls first;
endogenous Y1 X1 X3 X1X3;
instruments X2 X4 X2X4;
model Y1= X1 X3 X1X3;
run;

SAS PROGRAM FOR KENNY AND JUDD EMPIRICAL DATA (N=1160)
data a (type=cov);
infile cards missover;
input _name_ $ _type_ $ c1 v1 v2 s ssq v1s v2s v1ssq v2ssq;
cards;
c1    cov  2.626
v1    cov   .615  3.729
v2    cov  -.105  1.084  3.834
s     cov  -.724   .416   .358  4.963
ssq   cov   .252  -.629  -.182 -7.315  53.924
v1s   cov  2.317  -.677 -1.242  -.629   4.071  21.080
v2s   cov  1.321 -1.242 -.0685  -.183   1.507   7.577  23.545
v1ssq cov   .559 21.253  7.726  6.135 -25.205 -41.522 -28.734 381.402
v2ssq cov -4.015  7.725 23.673  3.284  -6.781 -28.884 -49.602 181.204 460.865
n n 1160 1160 1160 1160 1160 1160 1160 1160 1160
;
run;
proc print;
proc syslin data=a 2sls first;
endogenous c1 v1 v1s v1ssq;
instruments v2 v2s v2ssq s ssq;
eta1: model c1 = v1 v1s v1ssq s ssq;
run;

SAS PROGRAM FOR ANOMIA EMPIRICAL DATA (N=1137)
/* reading in the data */
options nocenter;
```

```
libname in1 '~/work/gss/';
data a; set in1.gss7294;
if year ne 76 then delete;

/* creating and changing the variables */
data b; set a;
anomia=anomia2+anomia5+anomia6+anomia7+anomia9;
income=income/1000;
age2=age*age;
anomia=anomia/5;

/* creating the instrumental variables */
data c; set b;
helpconc=helpful*conclerg;
helpcone=helpful*coneduc;
helpconf=helpful*confed;
fairconc=fair*conclerg;
faircone=fair*coneduc;
fairconf=fair*confed;
trstconl=trust*conlegis;

/* performing the 2SLS, including the overidentification test */
proc syslin 2sls first;
endogenous anomia trust conlegis trstconl;
instruments helpful fair conclerg coneduc confed
            helpconc helpcone helpconf
            fairconc faircone fairconf age age2 female black other
            educ income cath;
model anomia=trust conlegis trstconl age age2 female black other
            educ income cath /overid; run;
```

7

Modeling Interaction Effects: A Two-Stage Least Squares Example

Fuzhong Li
Oregon Research Institute, Eugene

Peter Harmer
Willamette University, Salem, OR

Researchers in the social sciences often deal with complex models involving nonlinear relationships, such as interactions between predictor variables. The interaction effects operate when two independent variables influence each other so that the effect of either one on a dependent, or criterion variable depends on the level of the other. When the effects of continuous variables are analyzed in the context of multiple regression models, the standard ordinary least squares (OLS) regression approach is often adopted. However, measurement reliability in testing regression coefficients using traditional analyses has been a major concern in the literature (e.g., Aiken & West, 1991; Busemeyer & Jones, 1983; Jaccard, Turrisi, & Wan, 1990). Advances in methodology have allowed estimation of interaction effects within the framework of structural equation modeling (SEM), with the major advantage being able to account for measurement error. Recent examples include the use of two classes of estimation methods: (a) full-information maximum likelihood (ML; e.g., Jaccard & Wan, 1995; Jöreskog & Yang, 1996; Ping, 1996) and (b) limited-information two-stage least squares (2SLS; Bollen, 1995). Unfortunately, these procedures have not been widely applied in social science research. One reason for this may be the lack of concrete illustrations of the procedures in applied settings.

Focusing on the limited-information estimation procedure, the purpose of this chapter is to provide a substantive illustration of how to estimate structural equation models (SEMs) with interactions of latent variables using 2SLS. First, the use of SEM methodologies to test interaction effects is briefly reviewed. Then, a substantive example of an interaction model using sport motivation data is presented. Finally, a specific model concerning the effects of competence and autonomy on sport motivation is formulated and tested using 2SLS.

TESTING INTERACTION EFFECTS OF LATENT VARIABLES IN STRUCTURAL EQUATION MODELS

Interaction effects are typically evaluated by forming a product term between two independent variables in a system equation. However, when constructs are measured through multiple observed indicators, the analytic procedures are not as straightforward as conventional multiple regression analysis where all variables are observed (e.g., Aiken & West, 1991). Busemeyer and Jones (1983) and Kenny and Judd (1984) proposed methods that allow estimation of nonlinear effects (e.g., interactions, curvilinear effects) among latent variables using SEM methodology. There are, however, a number of limitations associated with these methods. For example, they do not provide tests of statistical significance of parameter estimates nor estimation of equation intercepts, thereby limiting their utility. Readers are referred to Bollen (1989, 1995) for a discussion of these methods and limitations.

Several extensions of Kenny and Judd's (1984) approach using the ML estimation procedure have been proposed in the past few years. These include the multiple-product-indicator approach (i.e., Jaccard & Wan, 1995; Ping, 1996) and a single-product-indicator approach (Jöreskog & Yang, 1996). The multiple-product-indicator approach requires centering of the raw scores and multivariate normality. The single-product-indicator approach, also assuming multivariate normality, makes use of raw data and incorporates mean structures, thereby allowing estimation of intercept terms in the system equation. Moreover, the approach simplifies the number of cross-product terms and nonlinear constraints by using only a single product variable for model identification. This method is considered superior to multiple-indicator-based methods in that it provides more sound parameter estimates in the general case of noncentered data (Jaccard & Wan, 1996). Li et al. (1998) provided an exposition of these ML-based methods using a general data set.

Adopting a different estimation approach, Bollen (1995, 1996) proposed a 2SLS estimator as a way of modeling latent variable product terms that can be applied to general SEMs (e.g., the latent variable and the measurement model). The method requires the use of instrumental variables and large sample sizes. Compared to ML-based methods, Bollen (1995) and Bollen and Paxton in chapter 6 noted that 2SLS offers an estimator that is simple and easy to implement, and has known asymptotic properties that do not depend on the normality of the observed random variables. Given these positive features, a demonstration of the utility of this method is useful for applied researchers who are interested in latent variable product term analysis. The next section presents a theoretical model derived from the sport motivation literature and uses it as a substantive example in the further demonstration of the 2SLS approach.

A SUBSTANTIVE EXAMPLE OF AN INTERACTION MODEL

Testing interaction effects of latent variables in SEMs requires a sound theoretical justification (Jöreskog & Yang, 1996). For the purpose of this chapter, the self-determination theory proposed by Deci and Ryan (1985) is used to illustrate the application of the 2SLS technique. The theory proposes that motivation for engaging in goal-directed activities varies with changes in perceptions of competence and self-determination/autonomy. Although a wealth of empirical research supports the basic premise that perceptions of competence and autonomy influence levels of motivation (e.g., Brière, Vallerand, Blais, & Pelletier, 1996; Li, 1996; Pelletier et al., 1995), one critical proposition of the theory that has been ignored in most empirical investigations is the effect of the interaction of competence and autonomy on intrinsic motivation. As proposed by Deci and Ryan (1985), an increase in the perceptions of competence enhances intrinsic motivation only when accompanied by feelings of autonomy. Thus, the relation between one's competence in doing an activity (e.g., weight lifting) and the intrinsic motive for doing that activity should be moderated by feelings of autonomy. Although studies have provided indirect support for this proposition (e.g., Fisher, 1978; Markland & Hardy, 1997; Ryan, 1982), the interaction effect has not been tested directly. In the next section, a description of the data used for testing this interaction and an outline of the model parameter specifications in 2SLS analysis are provided.

METHOD

Participants and Procedure

The sample consisted of 869 university students (men, n = 436, M age = 20.67 years, SD = 2.54; women, n = 433, M age = 20.92 years, SD = 2.92). Participants were enrolled in various physical activity skills classes, including basketball, volleyball, martial arts, fencing, weight training, and swimming. Data were collected during scheduled class time in the regular school year.

Measures

Perceptions of Sport Competence. The sports competence subscale of the Physical Self-Perception Profile (Fox, 1990) was used to measure perceptions of competence. The subscale contains six items in a 4-point structured-alternative response format, with a score of 4 indicative of high perceived competence. Confirmatory factor analysis (CFA) was used to assess the internal consistency of the measure with multi-item reflective indicators. Model fitting of the single latent construct of competence indicated an acceptable fit of the model to the data; $\chi^2(9, N = 869) = 23.573, p < .01$, NNFI = .969, RMSEA = .073. Factor loadings ranged from .918 to .927 with an average of .925 (standardized). Further testing on a model of tau-equivalent measures (i.e., the patterns of factor loadings being equal: $\lambda_1 = \lambda_2 = \lambda_3 = \lambda_4 = \lambda_5 = \lambda_6$) indicated equivalent items loading on the hypothesized single construct, $\chi^2(14, N = 869) = 29.343, p < .01$, NNFI = .978, RMSEA = .069. The difference in chi-square test, $\chi^2(5, N = 869) = 29.343 - 23.573 = 5.770, p = .33$, indicated no significant deterioration given the equality constraints. All six items of the instrument showed tau-equivalent measures of the same underlying Competence construct. For ease of presenting the 2SLS procedure, two three-item averaged scores were constructed via random splitting of the measure and used as indicators in defining the Competence construct.

Perceptions of Exercise Autonomy. A modified Internality subscale of the Exercise Objectives Locus of Control Scale (McCready & Long, 1985) was used to assess perceptions of autonomy in sport. The Internality subscale consists of six items anchored on a 5-point Likert scale, ranging from (1) Strongly disagree to (5) Strongly agree. CFA showed an acceptable fit of the model to the data, $\chi^2(9, N = 869) = 45.825, p < .001$, NNFI = .963, RMSEA = .040. Factor loadings ranged from .974 to .982, with an average loading of .974 (standardized). A test of the tau-equivalent model indicated the items were equivalent in assessing the single construct of autonomy, $\chi^2(14, N = 869) = 49.773, p < .001$, NNFI = .954, RMSEA = .064. The chi-square difference test between the two models was not significant, $\chi^2(5, N = 869) = 3.948, p = .56$, indicating the hypothesis of congeneric measures was tenable. As a result, two three-item averaged scores were used to define the construct of Autonomy for the same reason described earlier.

Intrinsic Motivation. The Intrinsic Motivation (IM) measure was taken from the Sport Motivation Scale developed by Pelletier et al. (1995) and used as a single indicator of the IM construct in this analysis. The IM measure consists of three subscales: IM to know, IM to accomplish, and IM to experience stimulation. Each of these subscales consists of four items utilizing a 7-point scale, ranging from (1) Strongly disagree to (7) Strongly agree. A CFA was conducted to examine the single-factor model of IM.

In the CFA, scores from the three subscales were used as indicators of the IM construct. With three measures, the CFA model is exactly identified. An exactly identified model provides enough information to estimate all parameters, but, because degrees of freedom (*df*) equal zero for the model, the chi-square test indicates a perfect fit. To test how well the single-factor model with three measures fits the data, all three factor loadings (λ_{11}, λ_{21}, and λ_{31}) were constrained to be equal. This yielded a chi-square value of 5.636, $p = .06$, with 2 *df.* The test indicated that the three subscale scores of the IM each loaded significantly and equally on the IM construct. Therefore, using a composite score in the interaction model analysis was justified.

The Interaction Model of Intrinsic Motivation

An operationalized form of the interaction model to be estimated is presented in Fig. 7.1. On the left side of the model, all constructs are defined by two observed indicators to form latent variables. x_1 and x_2 are indicators of a latent variable L_1 for Competence, and x_3 and x_4 are indicators for a latent

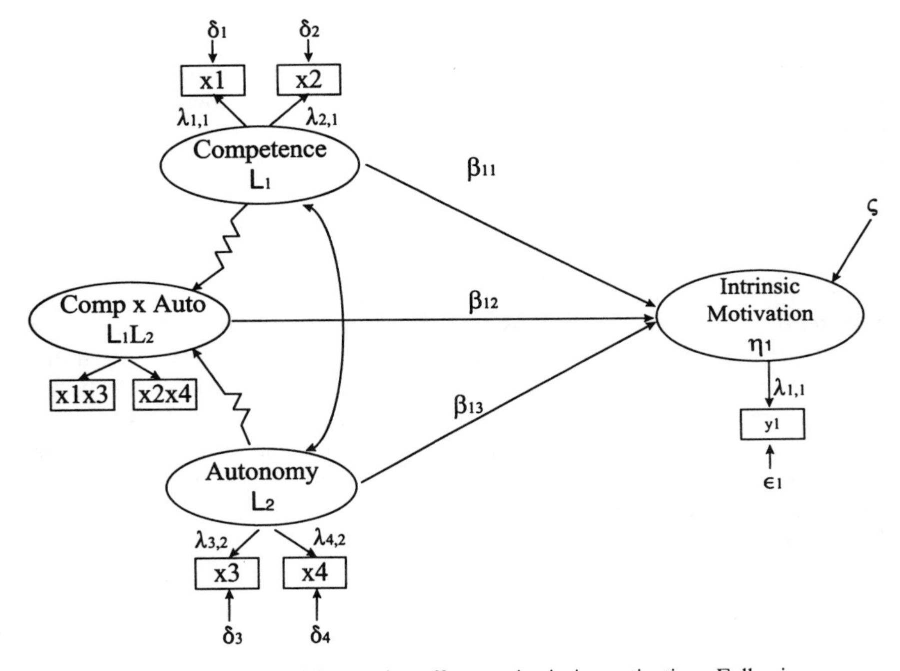

FIG. 7.1. Model of interactive effect on intrinsic motivation. Following Bollen (1995) and Bollen and Paxton (chap. 6), the saw-tooth arrows stand for interactive relationships between the variables at the base and the head of the arrow. Remaining symbols follow conventional SEM practice.

variable L_2 for Autonomy. A single indicator, y_1 (composite score), was used for the latent variable η_1, IM.

As theorized, the model suggests an interaction effect (L_1L_2) of perceptions of Competence (L_1) and Autonomy (L_2) on IM (η_1), in addition to the independent effects of Competence and Autonomy. That is, beyond the independent effects, high levels of competence in doing an activity have a positive effect on intrinsic motivation for individuals who perceive some self-determination. Expressing the structural relationships specified in the interaction model with a multiple regression equation obtains the following:

$$\eta_1(\text{IM}) = \alpha + \beta_{11}L_1(\text{Comp}) + \beta_{12}L_2(\text{Auto}) + \beta_{13}L_1L_2(\text{Comp} \times \text{Auto}) + \zeta \quad (1)$$

where η_1 is the dependent (outcome) variable of IM, α is the intercept term, β_{11} and β_{12} (representing the main effects of L_1 and L_2, respectively) and β_{13} (representing the interaction effect of L_1L_2) are population regression coefficients for the independent variables, and ζ is a disturbance term in the equation.

Model Specifications and Estimation Using 2SLS

Like other SEM-based methods, products of observed variables must be formed in applying the 2SLS method. However, they are treated as instrumental variables (IVs) to be used in the first-stage regression analysis. To facilitate the description of the 2SLS application, Equation 1 is repeated here:

$$\eta_1 = \alpha + \beta_{11}L_1 + \beta_{12}L_2 + \beta_{13}L_1L_2 + \zeta, \quad (2)$$

with the fundamental assumptions of ζ (a) having a zero mean (i.e., $E(\zeta) = 0$), and (b) being identically and independently distributed, (c) being independent of L_i for $i = 1,2$, and (d) being nonautocorrelated across observations.

Also, as with any SEM application, the origin and the unit of measurement of each latent variable must be defined. 2SLS requires that the latent variable in the equation be scaled to one of its indicators and the model equation be expressed in observed form (Bollen, 1995). In the model shown in Fig. 7.1, the scaling indicators for variables L_1 and L_2 are: $x_1 = L_1 + \delta_1$ or $L_1 = x_1 - \delta_1$ for the latent variable of Competence, and $x_3 = L_2 + \delta_3$ or $L_2 = x_3 - \delta_3$ for the latent variable of Autonomy. The scaling indicator assigned for L_1L_2 is a product term of x_1 times x_3:

$$L_1L_2 = (x_1 - \delta_1)(x_3 - \delta_3)$$
$$= x_1x_3 - x_1\delta_3 - x_3\delta_1 + \delta_1\delta_3$$

for the latent interaction variable of Competence × Autonomy. The nonlinear function of the latent variable L_1L_2 is decomposed into two components: x_1x_3 and $-x_1\delta_3 - x_3\delta_1 + \delta_1\delta_3$ (see Bollen, 1995, Equations 19 and 20). The first term, x_1x_3, is used in the main part of the model (i.e., defining the product L_1L_2) whereas the latter terms are considered residuals and placed in the composite disturbance term u introduced later. Finally,

$$y_1 = \eta_1 + \varepsilon_1 \text{ or } \eta_1 = y_1 - \varepsilon_1$$

for the latent outcome variable of IM.

Equation 2 is replaced using the observed scaling variables outlined previously. This is done by placing the scaling indicators x_1, x_3, x_1x_3 (i.e., $x_1 - \delta_1$ for L_1, $x_3 - \delta_3$ for L_2, and x_1x_3 for L_1L_2 variable) into Equation 2, which leads to the following equation:

$$y_1 = \alpha + \beta_{11}x_1 + \beta_{12}x_3 + \beta_{13}x_1x_3 + u_1 \tag{3}$$

where $u_1 = -\beta_{11}\delta_1 - \beta_{12}\delta_3 - \beta_{13}x_1\delta_3 - \beta_{13}x_3\delta_1 + \beta_{13}\delta_1\delta_3 + \varepsilon_1 + \zeta$. The u_1 term specified here is a composite disturbance term that contains δ_1, δ_3, $-x_1\delta_3 - x_3\delta_1 + \delta_1\delta_3$, ε_1, and ζ. These additional error terms are derived from the substitution of the observed scaling indicator for its underlying latent variable. The problem with estimating Equation 3 with OLS is that the composite disturbance term u_1 is correlated with some of the right-hand-side variables. For example, it can be seen that u_1 contains δ_1 and δ_3, the error of measurement for indicators of x_1 and x_3. This violates the assumption of OLS that the regressors are uncorrelated with the error term. As a result, the estimates from OLS regression are biased. This problem can be resolved by using the 2SLS estimator, which uses the IV method for arriving at consistent parameter estimates.

Selection of Instrumental Variables. Implementation of the 2SLS technique requires creation and selection of a set of IVs for the model estimation (Bollen, 1995). The basic assumption underlying the selection of IVs is that these IVs are correlated with the right-hand-side variables in Equation 3, but are not correlated with any component in the composite u_1. Bollen (1989, 1995) and Bollen and Paxton in chapter 6 discuss in detail how IVs should be selected, and suggest inspection of the covariance between a potential IV variable, x_i, and the composite disturbance, u, of a model to determine the eligibility of a potential IV.

The covariance between x_i, where x_i is a possible IV, and u can be calculated by multiplying the two, taking expectations, and using the assumptions that errors and factors are uncorrelated, and that pairs of errors are uncorrelated. For example, taking x_2 in Fig. 7.1 as a potential IV, the covariance between x_2 and u_1 is given as:

$$\begin{aligned}\mathrm{Cov}(x_2,u_1) = {} & \mathrm{Cov}(\lambda_{21}L_1 + \delta_2, -\beta_{11}\delta_1 - \beta_{12}\delta_3 - \beta_{13}x_1\delta_3 - \beta_{13}x_3\delta_1 + \beta_{13}\delta_1\delta_3 + \varepsilon_1 + \zeta) \\ = {} & \mathrm{Cov}(\lambda_{21}L_1, -\beta_{11}\delta_1) + \mathrm{Cov}(\lambda_{21}L_1, -\beta_{12}\delta_3) + \mathrm{Cov}(\lambda_{21}L_1, -\beta_{13}x_1\delta_3) + \\ & \mathrm{Cov}(\lambda_{21}L_1, -\beta_{13}x_3\delta_1) + \mathrm{Cov}(\lambda_{21}L_1, \beta_{13}\delta_1\delta_3) + \mathrm{Cov}(\lambda_{21}L_1, \varepsilon_1) + \\ & \mathrm{Cov}(\lambda_{21}L_1, \zeta) + \mathrm{Cov}(\delta_2, -\beta_{11}\delta_1) + \mathrm{Cov}(\delta_2, -\beta_{12}\delta_3) + \\ & \mathrm{Cov}(\delta_2, -\beta_{13}x_1\delta_3) + \mathrm{Cov}(\delta_2, -\beta_{13}x_3\delta_1) + \mathrm{Cov}(\delta_2, \beta_{13}\delta_1\delta_3) + \\ & \mathrm{Cov}(\delta_2, \varepsilon_1) + \mathrm{Cov}(\delta_2, \zeta).\end{aligned}$$

Substitute $(L_1 + \delta_1)$ and $(L_2 + \delta_3)$ for x_1 and x_3 in u_1 to obtain the following:

$$\begin{aligned}\mathrm{Cov}(x_2,u_1) = {} & -\lambda_{21}\beta_{11}\mathrm{Cov}(L_1,\delta_1) - \lambda_{21}\beta_{12}\mathrm{Cov}(L_1,\delta_3) - \lambda_{21}\beta_{13}\mathrm{Cov}(L_1, L_1\delta_3 + \delta_1\delta_3) - \\ & \lambda_{21}\beta_{13}\mathrm{Cov}(L_1, L_2\delta_1 + \delta_1\delta_3) + \lambda_{21}\beta_{13}\mathrm{Cov}(L_1, \delta_1\delta_3) + \lambda_{21}\mathrm{Cov}(L_1, \varepsilon_1) + \\ & \lambda_{21}\mathrm{Cov}(L_1, \zeta) - \beta_{11}\mathrm{Cov}(\delta_2, \delta_1) - \beta_{12}\mathrm{Cov}(\delta_2, \delta_3) - \beta_{13}\mathrm{Cov}(\delta_2, L_1\delta_3 + \delta_1\delta_3) - \\ & \beta_{13}\mathrm{Cov}(\delta_2, L_2\delta_1 + \delta_1\delta_3) + \beta_{13}\mathrm{Cov}(\delta_2, \delta_1\delta_3) + \beta_{13}\mathrm{Cov}(\delta_2, \varepsilon_1) + \mathrm{Cov}(\delta_2, \zeta).\end{aligned}$$

Simplify the preceding equation by removing covariances that are zero, under the assumptions that (a) L_1 is independent of δ_1, δ_3, ε_1, and ζ, and (b) δ_2 is independent of δ_1, δ_3, ε_1, and ζ.

$$\begin{aligned}\mathrm{Cov}(x_2,u_1) = {} & -\lambda_{21}\beta_{13}\mathrm{Cov}(L_1, L_1\delta_3 + \delta_1\delta_3) - \lambda_{21}\beta_{13}\mathrm{Cov}(L_1, L_2\delta_1 + \delta_1\delta_3) + \lambda_{21}\beta_{13}\mathrm{Cov}(L_1, \delta_1\delta_3) - \\ & \beta_{13}\mathrm{Cov}(\delta_2, L_1\delta_3 + \delta_1\delta_3) - \beta_{13}\mathrm{Cov}(\delta_2, L_2\delta_1 + \delta_1\delta_3) + \beta_{13}\mathrm{Cov}(\delta_2, \delta_1\delta_3).\end{aligned}$$

Take the expectations, using the equation of $\mathrm{Cov}(x,y) = E(xy) - E(x)E(y)$.

$$\begin{aligned}\mathrm{Cov}(x_2,u_1) = {} & -\lambda_{21}\beta_{13}[\mathrm{E}(L_1^2\delta_3) - E(L_1)E(L_1\delta_3) + E(L_1\delta_1\delta_3) - E(L_1)E(\delta_1\delta_3)] - \\ & \lambda_{21}\beta_{13}[E(L_1L_2\delta_1) - E(L_1)E(L_2\delta_1) + E(L_1\delta_1\delta_3) - E(L_1)E(\delta_1\delta_3)] + \\ & \lambda_{21}\beta_{13}[E(L_1\delta_1\delta_3) - E(L_1)E(\delta_1\delta_3)] - \\ & \beta_{13}[E(\delta_2L_1\delta_3) - E(\delta_2)E(L_1\delta_3) + E(\delta_2\delta_1\delta_3) - E(\delta_2)E(\delta_1\delta_3)] - \\ & \beta_{13}[E(\delta_2L_2\delta_1) - E(\delta_2)E(L_2\delta_1) + E(\delta_2\delta_1\delta_3) - E(\delta_2)E(\delta_1\delta_3)] + \\ & \beta_{13}[E(\delta_2\delta_1\delta_3) - E(\delta_2)E(\delta_1\delta_3)].\end{aligned}$$

Using the aforementioned assumptions (a) and (b) and the assumption that $E(L_1) = E(L_2) = E(\delta_1) = E(\delta_2) = E(\delta_3) = 0$, then $\mathrm{Cov}(x_2,u_1) = 0$. Therefore, x_2 qualifies as an IV. A similar procedure can be applied to determine IVs that are nonlinear functions of the observed variables. Bollen and Paxton in chapter 6 also suggest that the products of indicators that are not used for scaling the respective latent variables are often candidates for use as IVs. The product variable x_2x_4 in Fig. 7.1 meets this condition. The product variable x_1x_3, used as a scaling indicator, is not eligible because the error of measurement for the scaling indicator is part of the u_1 disturbance term.

As a result, a set of three variables, x_2, x_4, and x_2x_4, are identified as IVs and are used to obtain predicted values for the endogenous variables (y_1, x_1, x_3, x_1x_3) by a first-stage regression. Indicators of x_1, x_3, and x_1x_3 are naturally excluded because of their correlations with the u_1 term. The model is estimated using 2SLS in the SAS Syslin subroutine (SAS, 1993). The first stage of 2SLS estimation uses IVs to predict the "endogenous" explanatory variables by OLS regression. In the case of multiple IVs, the 2SLS estimator is an IV estimator that uses an optimal combination of IVs (Bollen, 1995). In the second stage of 2SLS, the predicted values of the endogenous explanatory variables from the first-stage OLS regression are used to estimate the coefficients in the original system by OLS regression (i.e., Equation 3). Readers are referred to Bollen (1995, 1996) and Bollen and Paxton (chap. 6) for details on model assumptions and derivation of the 2SLS estimator. The SAS commands used for this chapter are supplied in the Appendix. Also included in the Appendix are SPSS (1990) commands for the 2SLS procedure.

RESULTS

The quality of IVs used in the model was examined by inspecting the squared multiple correlation coefficient (R^2). In practice, low values of R^2 (e.g., $< .10$) indicate that the IVs selected may not be suitable for the model even though they satisfy the conditions required of IVs (Bollen, 1995). The R-squares for the first-stage regressions of the 2SLS procedure for x_1, x_3, and x_1x_3 were .754, .635, and .717, respectively. These R-squares were high, indicating the IVs chosen for the model (x_2, x_4, and x_2x_4) worked well for the example data.

The estimated unstandardized path coefficients and their standard errors of estimates (SE) for Equation 1 using the 2SLS estimator were:

$$\begin{array}{llll} \text{IM} = \underset{(.514)}{.719} + \underset{(.181)}{.067}\,(\text{Competence}) & + \underset{(.158)}{-.003}\,(\text{Autonomy}) & + \underset{(.056)}{.277}\,(\text{C} \times \text{A}) \end{array}$$

where the estimates in parentheses represent standard error.

Results indicated support for the theoretical predictions. That is, a significant interaction effect (slope) emerged so that perceptions of competence interacted with autonomy to jointly influence intrinsic motivation, $t = 4.989$, $p < .001$. The two main effects were not statistically significant in the presence of the interaction effect: $t = .372$, $p = .780$ for the perceptions of competence and $t = -.019$, $p = .985$ for the perceptions of autonomy.

The effect size of the interaction, defined as the increment in squared multiple correlation (R^2) over and above a main effect (Jaccard et al., 1990), was assessed by taking the difference in R^2 for the interaction model (R^2y_1, $x_1\, x_3\, x_1x_3$) and the "main effect only" model (R^2y_1, $x_1\, x_3$).[1] For the interaction model, the R^2 was .408; for the latter, the R^2 was .367. The difference value of .041 indicates that the interaction effect accounts for approximately 4% of the variance in the indicator of IM. The effect size f^2 calculated using Cohen's (1988) formula[2] was .07, an effect size that is slightly larger than Cohen's small-size definition (i.e., .02), and a typically observed effect size in most psychological research (Champoux & Peters, 1987).

To further understand the influence of autonomy on the relationship between perceptions of competence and intrinsic motivation, additional analyses were conducted as recommended by Aiken and West (1991). This approach, termed *simple slope analysis,* involves creating the following two conditional values for autonomy: (a) high autonomy (scores above 50th percentile) and (b) low autonomy (scores below 50th percentile). These conditional autonomy values were used to form interactions (cross-product terms) with the competence scores, and two separate 2SLS regression models were specified to estimate the effects of perceptions of competence, perceptions of autonomy, and their interaction on intrinsic motivation. The slope analysis revealed that the effect of the competence and autonomy interaction on intrinsic motivation is strong when levels of autonomy are high (i.e., scores above 50th percentile), $\beta = .464$, $t = 3.195$, $p < .001$, and that the strength and significance of this relationship declines as levels of autonomy decrease (i.e., scores below 50th percentile), $\beta = .109$, $t = .588$, $p = .566$. This set of analyses suggests that physical activity participants who have high levels of self-determination with competence are more likely to report higher levels of intrinsic motivation in activity participation than are participants who report low levels of self-determination.

DISCUSSION

This chapter provided a demonstration of Bollen's 2SLS approach in estimating latent variable interaction effects in SEMs. Specifically, following

[1]The R^2's are derived from the second stage of 2SLS regression estimation and are defined as:

$$R^2 = \frac{RSS}{RSS + ESS}$$

where RSS = regression sum of squares, ESS = error sum of squares (SAS, 1993).

[2]The f^2 was calculated using the following formula:

$$\begin{aligned} f^2 &= (R^2y_1,\, x_1\, x_3\, x_1x_3 - R^2y_1,\, x_1\, x_3) \,/\, (1 - R^2y_1,\, x_1\, x_3\, x_1x_3) \\ &= (.408 - .367) \,/\, (1 - .408) \\ &= .069 \end{aligned}$$

where R^2 = Squared multiple correlation; y_1 = Intrinsic Motivation; x_1 = Competence; x_3 = Autonomy; x_1x_3 = Competence × Autonomy.

Bollen's (1995) work, it was shown how parameter estimates of an interaction model using 2SLS are specified in terms of model identifications (i.e., scaling latent variables and selecting IVs). Once the model is identified, estimation is relatively straightforward using the 2SLS estimator implemented by many common statistical packages (e.g., SAS, SPSS). Results from sport intrinsic motivation data demonstrated the utility of the 2SLS technique for interaction analysis involving latent variables. The methodology holds promise for explaining or accounting for interaction effects. Although a simple interaction model was illustrated in this chapter, other nonlinear models such as curvilinear effects can be estimated in a similar way. Bollen has provided further details of this procedure as well as examples (see chap. 6).

Some ML-based SEM procedures are not discussed in this chapter. Readers are referred to Bollen (1995) and Bollen and Paxton (chap. 6) for comparisons between the Kenny and Judd (1984) approach and 2SLS. Of particular interest is the recent single-product-indicator method proposed by Jöreskog and Yang (1996). Their method has been shown to be a simpler procedure compared to others (e.g., Jaccard & Wan, 1995; Ping, 1996) in that it minimizes the number of cross-product indicators that are commonly characterized by non-normal distributions. Some common features of Jöreskog and Yang's single indicator and Bollen's 2SLS method include (a) the use of raw data, (b) estimation of equation intercepts, and (c) requiring fewer products of observed variables. Because Jöreskog and Yang's method utilizes a ML-based estimation procedure, it rests on the assumption that the joint distribution of the observed variables is multivariate normal (although LISREL-WLS based on the augmented moment matrix provides an alternative—see Jöreskog & Yang, 1996; chap. 2, this volume). In contrast, the 2SLS procedure relaxes this assumption, providing an alternative solution to latent variable interaction analysis in situations where distributions are a major concern in the data. Additional advantages of the 2SLS over other SEM-based procedures include its ease of use and simple implementation (i.e., with conventional statistical software and straightforward multiple regression-like programming). These features are attractive to those who are comfortable with regression and/or analysis of variance models in testing interactions.

From an estimation perspective, the 2SLS is a limited-information method in that it estimates a single equation at a time and does not take into account complete information from all other structural equations in the model (Bollen, 1989). The ML-based single-indicator method, on the other hand, uses a full-information method of estimating parameters, which may lead to more efficient estimators. The limited-information estimation procedure may be practically useful in isolating specification error in a system equation or structural model (Bollen, 1995). A complete assessment of the relative merits of the limited-information 2SLS method and Jöreskog and Yang's (1996) single-indicator ML method awaits future research.

Some Implications for the Sport Motivation Model

The 2SLS analysis indicated that, consistent with the theory, the relationship between competence and intrinsic motivation is moderated by levels of autonomy. When levels of self-determination are high, the effect of competence on intrinsic motivation is strengthened, whereas when self-determination is low, the influence of competence on motivation is muted. Substantively, the findings of the study showed that an interactive function is operating in physical activity settings, suggesting that experiences of competence in engaging in an activity alone are not sufficient to promote intrinsic motivation. Perceptions of competence interact with perceptions of autonomy to jointly influence intrinsic motivation. Findings also suggested that research focusing only on the independent effects of competence and autonomy on intrinsic motivation are likely to provide a narrow understanding of motivation in goal-directed behavior.

CONCLUSIONS

With advances in SEM methodology, analytic procedures have become increasingly available for social science researchers to estimate complex models involving nonlinear relationships among latent variables. The 2SLS estimation procedure presented in this chapter is a practical example of how new techniques may be preferable to conventional analytic procedures, such as multiple regression with observed variables, in addressing interaction issues. In addition, compared to other SEM-based procedures (e.g., Jaccard & Wan, 1995; Jöreskog & Yang, 1996), the 2SLS offers a simpler procedure (i.e., computationally more simple) with fewer restrictions (e.g., distribution assumptions, mean centering, nonlinear constraints on variances) for testing interactions of latent variables in SEMs.

ACKNOWLEDGMENTS

The authors gratefully acknowledge assistance from Kenneth Bollen in the model specification necessary for the 2SLS analysis in this study. The authors would also like to thank Alan Acock and Lisa Strycker for their helpful comments on an earlier version of this chapter. Preparation of this chapter was supported in part by Grant DA 09306 from the National Institute on Drug Abuse.

REFERENCES

Aiken, L. S., & West, S. G. (1991). *Multiple regression: Testing and interpreting interactions.* Newbury Park, CA: Sage.

Bollen, K. A. (1989). *Structural equations with latent variables.* New York: Wiley.

Bollen, K. A. (1995). Structural equation models that are nonlinear in latent variables: A least-squares estimator. In P. M. Marsden (Ed.), *Sociological methodology* (pp. 223–252). Oxford, England: Blackwell.

Bollen, K. A. (1996). An alternative two stage least squares (2SLS) estimator for latent variable equations. *Psychometrika, 61,* 109–121.

Brière, N. M., Vallerand, R. J., Blais, M. R., & Pelletier, L. G. (1996). Development and validation of a measure of intrinsic, extrinsic, and amotivation in sports: The Sport Motivation Scale (SMS). *International Journal of Sport Psychology, 26,* 465–489.

Busemeyer, J. R., & Jones, L. E. (1983). Analysis of multiplicative combination rules when the causal variables are measured with error. *Psychological Bulletin, 93,* 549–562.

Champoux, J. E., & Peters, W. S. (1987). Form, effect size, and power in moderated regression analysis. *Journal of Occupational Research, 60,* 243–255.

Cohen, J. (1988). *Statistical power analysis for the behavioral sciences* (2nd ed.). Hillsdale, NJ: Lawrence Erlbaum Associates.

Deci, E. L., & Ryan, R. M. (1985). *Intrinsic motivation and self-determination in human behavior.* New York: Plenum.

Fisher, C. D. (1978). The effects of personal control, competence, and extrinsic reward systems on intrinsic motivation. *Organizational Behavior and Human Performance, 21,* 273–288.

Fox, K. R. (1990). *The physical self-perception profile manual.* DeKalb: Northern Illinois University Press.

Jaccard, J., Turrisi, R., & Wan, C. K. (1990). *Interaction effects in multiple regression* (Sage University Papers series on Quantitative Applications in the Social Sciences). Newbury Park, CA: Sage.

Jaccard, J., & Wan, C. K. (1995). Measurement error in the analysis of interaction effects between continuous predictors using multiple regression: Multiple indicator and structural equation approaches. *Psychological Bulletin, 117,* 348–357.

Jaccard, J., & Wan, C. K. (1996). *LISREL approaches to interaction effects in multiple regression* (Sage University Papers series on Quantitative Applications in the Social Sciences). Thousand Oaks, CA: Sage.

Jöreskog, K. G., & Yang, F. (1996). Non-linear structural equation models: The Kenny–Judd model with interaction effects. In G. A. Marcoulides & R. E. Schumacker (Eds.), *Advanced structural equation modeling: Issues and techniques* (pp. 57–88). Mahwah, NJ: Lawrence Erlbaum Associates.

Kenny, D., & Judd, C. M. (1984). Estimating the nonlinear and interaction effects of latent variables. *Psychological Bulletin, 96,* 201–210.

Li, F. (1996). *The Exercise Motivation Scale: Its multifaceted structure and construct validity.* Unpublished doctoral dissertation, Oregon Stage University, Corvallis.

Li, F., Harmer, P., Duncan, T. E., Duncan, S. C., Acock, A., & Boles, S. (1998). Approaches to testing interaction effects using structural equation modeling methodology. *Multivariate Behavioral Research, 33*(1), 1–40.

Markland, D., & Hardy, L. (1997). On the factorial and construct validity of the intrinsic motivation inventory: Conceptual and operational concerns. *Research Quarterly for Exercise and Sport, 68,* 20–32.

McCready, M. L., & Long, B. C. (1985). Locus of control, attitudes toward physical activity, and exercise adherence. *Journal of Sport Psychology, 7,* 346–359.

Pelletier, L. G., Fortier, M. S., Vallerand, R. J., Tuson, K. M., Brière, N. M., & Blais, M. R. (1995). Toward a new measure of intrinsic motivation, extrinsic motivation, and amotivation in sports: The Sport Motivation Scale (SMS). *Journal of Sport & Exercise Psychology, 17,* 35–53.

Ping, R. A. (1996). Latent variable interaction and quadratic effect estimation: A two-step technique using structural equation analysis. *Psychological Bulletin, 119,* 166–175.

Ryan, R. M. (1982). Control and information in the intrapersonal sphere: An extension of cognitive evaluation theory. *Journal of Personality and Social Psychology, 43,* 450–461.
SAS Institute Inc. (1993). *SAS/ETS: User's guide.* Cary, NC: Author.
SPSS Inc. (1990). *SPSS reference guide.* Chicago: Author.

APPENDIX

The following SAS commands were used for the model estimation presented in the chapter. The commands for SPSS are also provided.

```
SAS commands for the 2SLS procedure:
Title "Using 2SLS estimator for testing interaction effects";
Data inter;
  Infile 'inter.dat';
  Input y1 x1 x2 x3 x4 x1x3 x2x4;
Proc syslin 2sls first;
Instruments x2 x4 x2x4;
Endogenous x1 x3 x1x3 y1;
Model y1 = x1 x3 x1x3;
Run;

SPSS commands for the 2SLS procedure:
Title "Using 2SLS estimator for testing interaction effects"
Data list free file = "inter.dat"
  /y1 x1 x2 x3 x4 x1x3 x2x4
2SLS equation = y1 with x1 x3 x1x3
  /Instruments = x2 x4 x2x4
  /Endogenous = x1 x3 x1x3 y1
Finish
```

8

Effect Decomposition in Interaction and Nonlinear Models

Richard L. Tate
Florida State University

Single-equation modeling in the behavioral and social sciences (e.g., factorial analysis of variance and multiple regression) has long allowed the detection and description of functional complexities in relationships. For example, interactive regression models have permitted the description of relationships in which the effect of an independent variable on an outcome depends on the level of a second independent variable (often called a moderator variable; e.g., Aiken & West, 1991; J. Cohen & P. Cohen, 1983; Jaccard, Turrisi, & Wan, 1990; Saunders, 1956). Each of the two independent variables involved in an interaction can be either interval or categorical. Such interactions are commonly represented with sets of one or more product terms based on the two interacting variables (see Stolzenberg, 1980, for other formulations), and the interaction effect is then tested by testing the increase in R^2 due to adding the set of interaction variables. For an independent variable involved in an interaction, the effect on the outcome is the first partial derivative of the regression equation with respect to the variable of interest (e.g., Marsden, 1981; Stolzenberg, 1980). The result can be communicated to a reader by plotting the effect of an independent variable versus the other variable involved in the interaction and superimposing a confidence band on the resulting straight line (e.g., Tate, 1984).

Another common type of functional complexity modeled easily by single-equation procedures is a nonlinear relationship between two interval variables. Other mathematical forms are possible, but curvilinear relationships are often represented with polynomial models consisting of powers of the independent variable (e.g., J. Cohen & P. Cohen, 1983). The higher order terms in such a model are tested with standard techniques. In practice, the resulting relationship is usually described by plotting the predicted outcome versus the independent variable. In principle though, a formal expression for the effect of the independent variable on the outcome can be obtained by taking the first-order derivative of the regression equation with respect to the independent variable. This effect can be plotted versus the independent variable, superimposing a confidence band.

More recently there has been increasing interest in extending structural equation modeling (SEM) procedures to allow interactions and nonlinearities. The difficulty of model parameter estimation for interactive/nonlinear SEM depends dramatically on the generality of the model. For example, parameter estimation for recursive and nonrecursive models for observed variables is relatively simple. The structural equation for each endogenous variable would be elaborated to include any required product terms for interactions and power terms for nonlinearities. Standard SEM software could then be used to estimate all model coefficients. (The model specification for this is illustrated later.) If the model for observed variables is recursive (i.e., has no causal feedback loops and/or correlated errors), the coefficients for each structural equation could also be estimated independently of coefficients in other equations with standard multiple regression software.

Parameter estimation for interactive/nonlinear SEM is much more problematic when latent variables are modeled. Procedures for the single-equation special case of SEM with latent variables (i.e., a regression equation with latent independent variables) have been proposed, for example, by Heise (1986), Jaccard and Wan (1995, 1996), and Ping (1996b). For the general SEM situation with multiple structural equations, estimation procedures have been presented and illustrated by, for example, Hayduk (1987), Jöreskog and Yang (1996), Kenny and Judd (1984), and Ping (1996a). At this time, a review of these existing treatments suggests (at least to this author) that estimation procedures for interactive/nonlinear SEM with latent variables are still in development, and that procedures that are (a) capable of handling reasonably large and complex interactive/nonlinear models and (b) accessible to the typical modeling analyst are only now becoming available.

In addition to the question of parameter estimation, a second important issue related to interactive/nonlinear SEM is the "effect decomposition" required to describe the direct, indirect, and total effects implied by a model. Surprisingly, even for SEM with observed variables there has been relatively little discussion and illustration of effect decomposition for interactive/nonlinear models. An excellent but somewhat limited discussion was provided for recursive models by Stolzenberg (1980). The total effects of exogenous variables on endogenous variables were estimated in a two-step procedure in which the equations for the endogenous variables were put in reduced form and partial derivatives of the equations were obtained. (An equivalent version of this procedure is illustrated later.) Unfortunately, perhaps due to the desire to use a simple model for illustration, Stolzenberg did not consider indirect and total effects of endogenous variables on endogenous variables. Also, the example did not illustrate the possibility of an inherent uncertainty in the knowledge of the indirect and total effects of exogenous variables on endogenous variables, an uncertainty that is illustrated here. The only other treatment of effect decomposition for interactive models that could be found was that by Lance (1988). The approach described by Stolzenberg was applied after residualizing the product term in an interactive structural equation.

The purpose of this chapter is to extend the suggestions of Stolzenberg (1980) to cover all effects of potential interest in an interactive/nonlinear recursive SEM model, add several other perspectives from the current SEM literature, and illustrate the approach with a reasonably complex recursive model for observed variables. It is argued that the same procedure would also be appropriate for recursive SEM with latent variables once the estimation problem mentioned earlier is solved. More tentatively, it is suggested that the approach may also work with nonrecursive models for observed variables.

PROCEDURE

This section provides a brief and mathematical summary of the procedure. For the first reading, some readers may choose to go directly to the illustration provided in the next section.

Model Specification and Estimation

Assume a model with p endogenous observed variables (ys), q exogenous observed variables (xs), and r product and/or power terms required to represent the functional complexities in the model. Using a modification of the notation of Jöreskog and Sörbom (1989), an interactive, nonlinear model can be specified as

$$\mathbf{y} = \boldsymbol{\alpha} + \mathbf{B}^*\mathbf{y} + \boldsymbol{\Gamma}^*\mathbf{x}^* + \boldsymbol{\zeta} \tag{1}$$

where $\mathbf{y}$ is the $p \times 1$ vector of observed endogenous ys, $\mathbf{x}^*$ is a $(q + r) \times 1$ vector containing the q observed exogenous xs and the r required product and power terms of the xs and ys, $\boldsymbol{\alpha}$ is the $p \times 1$ vector of intercepts for the y equations, $\mathbf{B}^*$ is the $p \times p$ matrix containing the coefficients of the y variables in the structural equations for the ys, $\boldsymbol{\Gamma}^*$ is the $p \times (q + r)$ matrix containing the coefficients of the x^* variables in the y structural equations, and ζ is the $p \times 1$ vector of residuals. Given the presence of possible interaction and power terms in the model, the elements of $\mathbf{B}^*$ and $\boldsymbol{\Gamma}^*$ will not always represent the direct effects among the *x*s and *y*s.

Maximum likelihood estimation could be used to obtain the estimated model coefficients using any SEM software. With the focus in this study only on point estimation of the model parameters, an assumption of multinormality of the variables is not required. Maximum likelihood estimates are consistent for non-normal distributions (e.g., Bollen, 1989).

Effect Decomposition and Description

Direct Effects. Define the elements of the $p \times p$ matrix $\mathbf{B}$ as the direct effects of the *y*s on the *y*s and define the elements of the $p \times q$ matrix $\boldsymbol{\Gamma}$ as the direct effects of the *x*s on the *y*s. Following the discussion in, for example, Marsden (1981) and Stolzenberg (1980), these direct effects are determined with first-order partial derivatives. For example, element (j,k) in $\mathbf{B}$ (i.e., the direct effect of y_k on y_j) would be the first-order derivative of y_j with respect to y_k in the *j*th equation of Equation 1, and element (j,k) in $\boldsymbol{\Gamma}$ (i.e., the direct effect of x_k on y_j) would be the first-order derivative of y_j with respect to x_k in the *j*th equation of Equation 1.

Indirect and Total Effects. For additive/linear models, the fundamental notion of an indirect effect resulting from the transmission of successive changes from variable to variable through a path of multiple links can be represented in matrix form as given in standard works (e.g., Bollen, 1987, 1989; Jöreskog & Sörbom, 1989). Because the same fundamental logic defining an indirect effect will also apply in the presence of interactions and nonlinearities, the same matrix formulation for indirect (and total) effects will be appropriate for interactive/nonlinear models. These expressions for the total effects of the *y*s on the *y*s (TE[*y* on *y*]) and of the *x*s on the *y*s (TE[*x* on *y*]) are expressed in terms of the matrices of direct effects as follows:

$$\mathbf{TE(y\ on\ y)} = \mathbf{\Sigma B^k},$$
$$\mathbf{TE(x\ on\ y)} = [\mathbf{I} + \mathbf{\Sigma B^k}]\boldsymbol{\Gamma}. \tag{2}$$

The summation of powers of $\mathbf{B}$ is for $k = 1$ to infinity. For recursive models, there is a finite number of non-null $\mathbf{B}$ powers.

The total indirect effect for a given pair of variables can then be found by subtracting the corresponding direct effect from the total effect. (It's possible to further decompose the indirect effects; e.g., see Fox [1985] for the computation of "specific" effects defined as indirect effects mediated by a specific set of endogenous variables.)

For recursive models, path tracing provides an alternative to the matrix-based approach in Equation 2. First, paths in a path diagram would be labeled with the derived direct effect expressions described earlier. Application of standard path-tracing rules would then result in expressions identical to those obtained from Equation 2. This approach would usually be the simplest for small recursive models with modest numbers of variables. However, as the number of variables in the model increases, it becomes increasingly difficult to recognize all legitimate paths and the determination of indirect and total effects based on Equation 2 becomes safer.

Description. The last step in the procedure is the likely use of graphical representations to describe the resulting multidimensional relationships. In some cases it will be necessary to convert effect expressions to reduced form, resulting in effects expressed in terms of only the xs and the y residuals. This is accomplished by substitution of the reduced equations for the ys into the effect expressions.

ILLUSTRATION

Consider a recursive model with three endogenous observed variables (y_1, y_2, and y_3) and two exogenous observed variables (x_1 and x_2) defined by the following structural equations:

$$\begin{aligned} y_1 &= .5x_1 + .3x_2 + .1x_1x_2 + \zeta_1, \\ y_2 &= .2x_2 - .2x_2{}^2 + .3y_1 + \zeta_2, \text{ and} \\ y_3 &= .3y_1 + .4y_2 + .2y_1y_2 + \zeta_3. \end{aligned} \tag{3}$$

In terms of the general model in Equation 1, $\mathbf{y}' = (y_1\ y_2\ y_3)$ and $\mathbf{x}^{*\prime} = (x_1\ x_2\ x_1x_2\ x_2^2\ y_1y_2)$. In this example, (a) x_1 and x_2 interact in determining y_1, (b) the relationship of y_2 with x_2 is quadratic in form, and (c) y_1 and y_2 interact in determining y_3.

For this illustration, the variables x_1 and x_2 were assumed to be standardized with mean zero and variance one. It is important to note that the third set of equations were *not* in deviation form because the various product and power terms were not deviation scores. Thus, each equation included a constant, assumed here for simplicity to be zero. The correlation between x_1 and x_2 was equal to 0.3, and the standard deviations (*SD*s) of the three residuals in the structural equations (ζs) were equal to 0.5, 0.4, and 0.4, respectively.

Simulated data ($n = 5{,}000$) were generated with the aforementioned model and analyzed (assuming the same model) with the LISREL 8 program (Jöreskog & Sörbom, 1989). As expected, the estimated model fit the observed covariances very well (e.g., RMSE = .00954). Also as expected, the generating values were recovered to a high degree of precision. Therefore, the single-digit generating values, not the multiple-digit estimates, were used for the illustration coming later. From the LISREL results, the R^2 values for the three endogenous variables were 0.64, 0.60, and 0.55, respectively. The standard deviations for the endogenous variables (used later to standardize effect sizes) were 0.83, 0.64, and 0.60, respectively.

Direct Effects

For the direct effects of the *y*s on the *y*s, the appropriate partial derivatives in Equation 3 result in the following:

$$\begin{aligned} \mathrm{DE}(y_1 \text{ on } y_3) &= .3 + .2y_2, \\ \mathrm{DE}(y_2 \text{ on } y_3) &= .4 + .2y_1, \\ \mathrm{DE}(y_1 \text{ on } y_2) &= .3. \end{aligned} \tag{4}$$

The expressions in Equation 4 are elements (3,1), (3,2), and (2,1), respectively, in the **B** matrix. A graphical representation of the first two would simply plot the DEs as a function of y_2 and y_1, respectively. An alternative reduced format is discussed later.

Similarly, using the appropriate partial derivatives, the nonzero direct effects of *x*s on *y*s are the following:

$$\begin{aligned} \mathrm{DE}(x_1 \text{ on } y_1) &= .5 + .1x_2, \\ \mathrm{DE}(x_2 \text{ on } y_1) &= .3 + .1x_1, \\ \mathrm{DE}(x_2 \text{ on } y_2) &= .2 - (2)(.2)x_2, \\ &= .2 - .4x_2. \end{aligned} \tag{5}$$

These expressions are elements (1,1), (1,2), and (2,2) in the Γ matrix. They could also be easily represented in graphical form.

One possible format for an associated path diagram is shown in Fig. 8.1. Only the original *x*s and *y*s are represented as variables in the diagram, and the paths are labeled with the direct effects in Equations 4 and 5. An alternative path diagram format (e.g., J. Cohen & P. Cohen, 1983) would represent interactions with arrows from variables to path coefficients. This alternative format becomes more awkward when it is necessary to represent nonlinear relationships (there would be an arrow from a variable to the path coefficient in the path for that same variable). Moreover, as the number of functional complexities increases, the alternative diagram format would quickly become very cluttered.

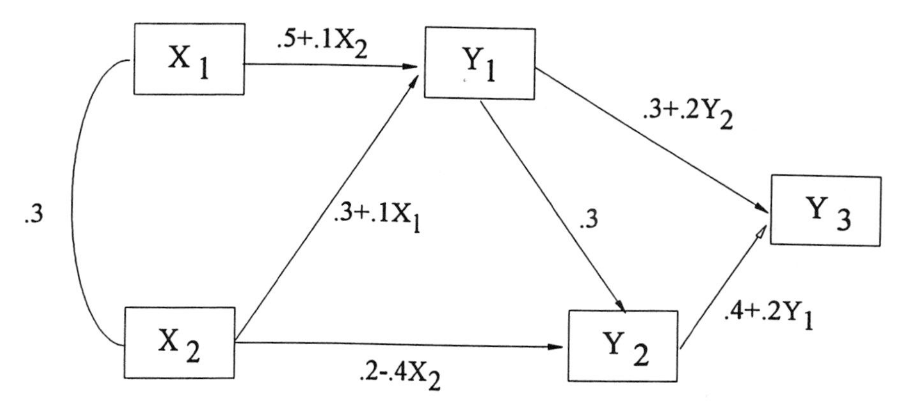

FIG. 8.1. Path diagram for the interactive/nonlinear example. Paths are labeled with the direct effect expressions that are obtained by taking first-order partial derivatives of structural equations.

Total and Indirect Effects

Effects of ys *on* ys. Determination of the total effects of ys on ys with the first expression of Equation 2 requires assessment of the powers of **B**. After squaring **B**, the only nonzero element (element [3,1]) is equal to (.3)(.4 + .2y_1). All higher powers of **B** are null. Thus, all total y on y effects are comprised only of the corresponding direct effects, except for the total effect of y_1 on y_3 which is equal to

$$\mathrm{TE}(y_1 \text{ on } y_3) = (.3 + .2y_2) + (.3)(.4 + .2y_1). \tag{6}$$

The second term in Equation 6 is the indirect effect of y_1 on y_3, mediated by y_2. This indirect effect depends on the level of y_1 because the effect of y_2 on y_3 depends on y_1. As indicated previously, this same result could also be obtained using standard path-tracing rules with Fig. 8.1.

One possible graphical format to describe the effect of y_1 on y_3 would plot the corresponding direct effect (first expression of Equation 4) and total effect (Equation 6) versus y_2. This is illustrated in Fig. 8.2. The effects have been scaled by standardizing with respect to the outcome variable (i.e., by dividing by the standard deviation of the outcome). As indicated in the first expression of Equation 4, the direct effect (DE) of y_1 on y_3 is a function only of the value of y_2. When the value of y_2 varies from 2 *SD*s below the mean to 2 *SD*s above (i.e., from −1.28 to +1.28), the direct effect of y_1 on y_3 varies from a very small positive value to approximately one, an appreciable change.

The total effect of y_1 on y_3 plotted versus y_2 is shown in Fig. 8.2 for two values of y_1, one 2 *SD*s above the mean and the other 2 *SD*s below. (The virtual identity of the direct effect and the total effect at the lower value of

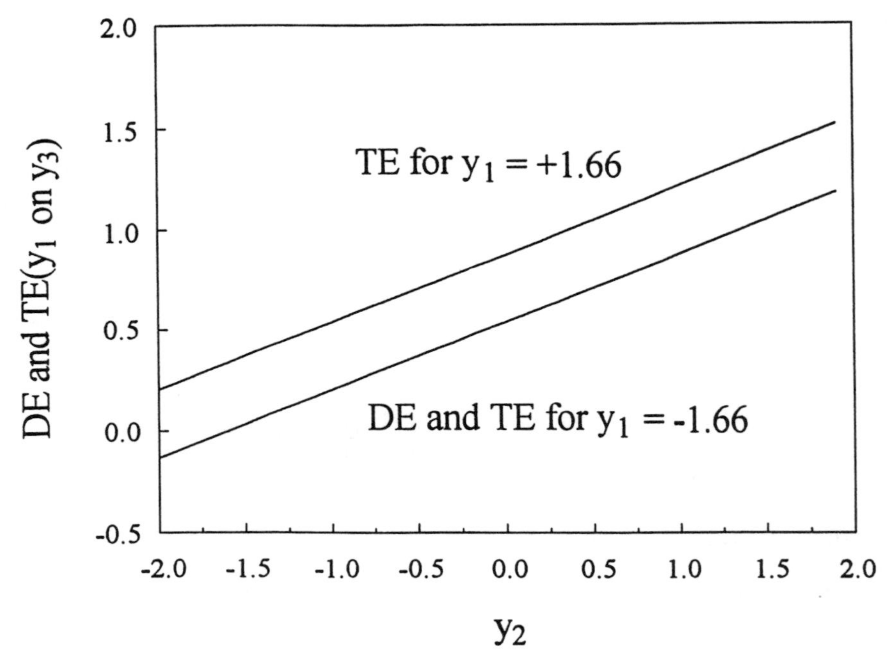

FIG. 8.2. The direct effect (DE) and total effect (TE) of y_1 on y_3 plotted as a function of y_2. The indirect effect, represented by the vertical separation between lines for the total and direct effects, depends on the level of y_1.

y_1 is coincidental.) The effect of y_2 on the total effect, holding constant y_1, is identical to that for the direct effect. Holding constant y_2, the effect of y_1 on the total effect is smaller than the effect of y_2 but not trivial. A plus and minus 2 *SD* change in y_1, holding y_2 constant, results in a change of approximately 0.3 units in the total effect.

The indirect effect of y_1 on y_3 is represented in Fig. 8.2 by the vertical separation of the lines representing the total and direct effects. The indirect effect is approximately zero when y_1 is 2 *SD*s below the mean and increases with increasing values of y_1, approaching a value between 0.3 to 0.4 for y_1 values 2 *SD*s above the mean.

The description offered in Fig. 8.2 is appropriate once specific values of y_1 and y_2 for an individual are considered known. It should be remembered, though, that the *y*s themselves are determined by the values of the exogenous variables. An alternative approach to the description of the effect of y_1 on y_3 would cast the description in reduced form, eliminating reference to values of the endogenous variables. If this were done, the resulting expression would have a systematic portion, expressed only in terms of values of *x*s, and a random component, involving random residuals of the endogenous variables. A description in this form is illustrated for another effect later.

Effects of xs *on* ys. The second expression of Equation 2 is used for the total effects of *x*s on *y*s. Alternatively, path tracing with the diagram in Fig. 8.1 will give the same results. To illustrate, consider the total effect of x_2 on y_3. Because there is no direct effect of x_2 on y_3, this total effect is entirely indirect, mediated by y_1 and y_2. After the operation in the second expression of Equation 2, element (3,2) is found to be

$$\begin{aligned}\text{TE}(x_2 \text{ on } y_3) = {} & (.3 + .1x_1)(.3 + .2y_2) + (.2 - .4x_2)(.4 + .2y_1) \\ & + (.3 + .1x_1)(.3)(.4 + .2y_1). \end{aligned} \tag{7}$$

An attempt to plot this relationship in this form would require a description of the effect as a function of four different variables. Such a description would be misleading because two of those variables, y_1 and y_2, depend on the other two variables, x_1 and x_2 (plus residual terms). Therefore, Equation 7 is cast in reduced form by substituting the reduced forms of the structural equations of y_1 and y_2. For the current illustration, the expression for y_1 in Equation 3 is already in reduced form and the reduced form of y_2 is found, by substitution of the first expression of Equation 3 into the second equation, to be

$$y_2 = .15x_1 + .29x_2 - .2x_2^2 + .03x_1x_2 + .3\zeta_1 + \zeta_2. \tag{8}$$

Substituting the reduced equations of y_1 and y_2 into Equation 7, the total effect of x_2 on y_3 can be expressed as the following:

$$\begin{aligned}\text{TE}(x_2 \text{ on } y_3) = {} & .206 + .08x_1 - .1252x_2 + .006x_1^2 \\ & - .036x_2^2 - .0248x_1x_2 - .012x_1x_2^2 + .0012x_1^2x_2 \\ & + [.076\zeta_1 + .06\zeta_2 + (.012\zeta_1 + .02\zeta_2)x_1 - .08\zeta_1x_2]. \end{aligned} \tag{9}$$

This same equation could also have been obtained using the approach of Stolzenberg (1980) by first obtaining the reduced equation for y_3, then determining the partial derivative of y_3 with respect to x_2.

Equation 9 reflects a total effect with both a systematic component depending only on the exogenous variables (the first two lines of Equation 9) and a random component resulting from the residuals for the endogenous variables (the expression in square brackets). The latter represents an inherent uncertainty about the total effect that is present even when the true model coefficients are known exactly. One possible way to describe the relationship in Equation 9 is to focus on the systematic and random components separately. That is, one would first show graphically the total effect of x_2 on y_3 for a person that is "typical" with respect to the random errors, that is, for a person with ζ_1 and ζ_2 both equal to 0. A possible three-dimensional (3-D) format for this is illustrated in Fig. 8.3. The surface shows the variation of the total effect,

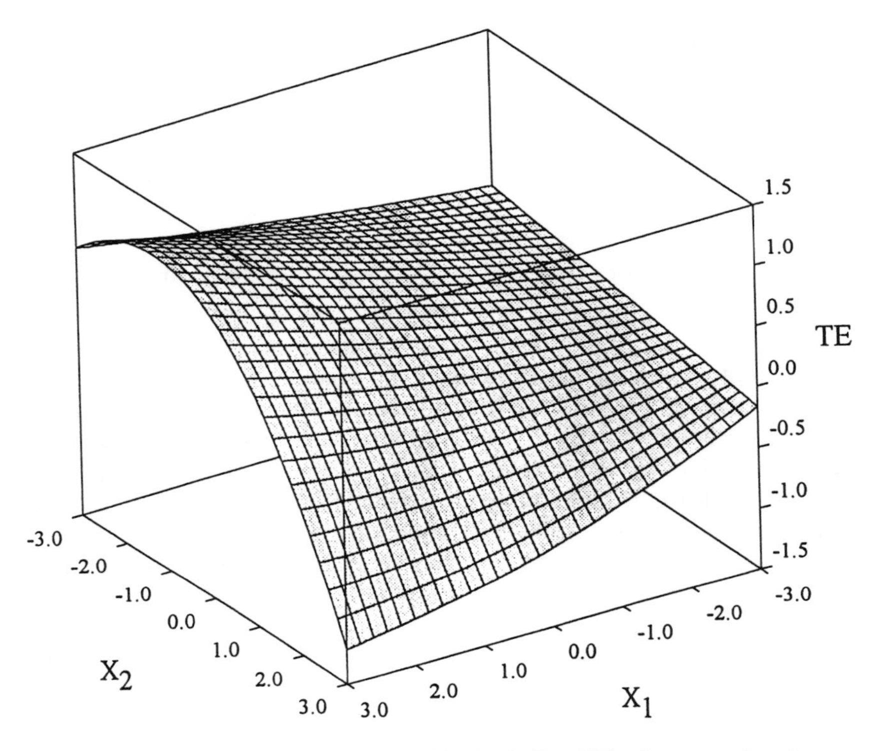

FIG. 8.3. The systematic portion of the total effect (TE) of x_2 on y_3 plotted versus x_1 and x_2. Given the absence of a direct effect of x_2 on y_3, this total effect is entirely indirect.

standardized with respect to y_3, as a function of x_1 and x_2. (Note that in my rotation of the graph to obtain the best view of the surface, the resulting directions of the x_1 and x_2 axes are the reverse of the usual.) The highest estimated positive effects, around 1.0, are found for a combination of very high x_1 values and low x_2 values. From this high point, the surface falls off very gradually in the x_1 direction and quite rapidly in the x_2 direction. For individuals with very low values on both x_1 and x_2, the total effect is quite small, whereas for those with very high values on both xs, the total effects are negative and quite strong. When x_1 is approximately equal to x_2 (remember there is a positive correlation between the xs), the total effect ranges from a modest positive value for low xs to a very strong negative value for high xs.

An alternative format based on constant effect contours is shown for the same results in Fig. 8.4. This contour format allows better quantification of some of the features discussed earlier for Fig. 8.3. Consider, for example, the discussion of the total effect when an individual's x_1 value is equal to their x_2 value. In Fig. 8.4, this constraint corresponds to the diagonal from the lower left of the figure to the upper right. For all x values between –2

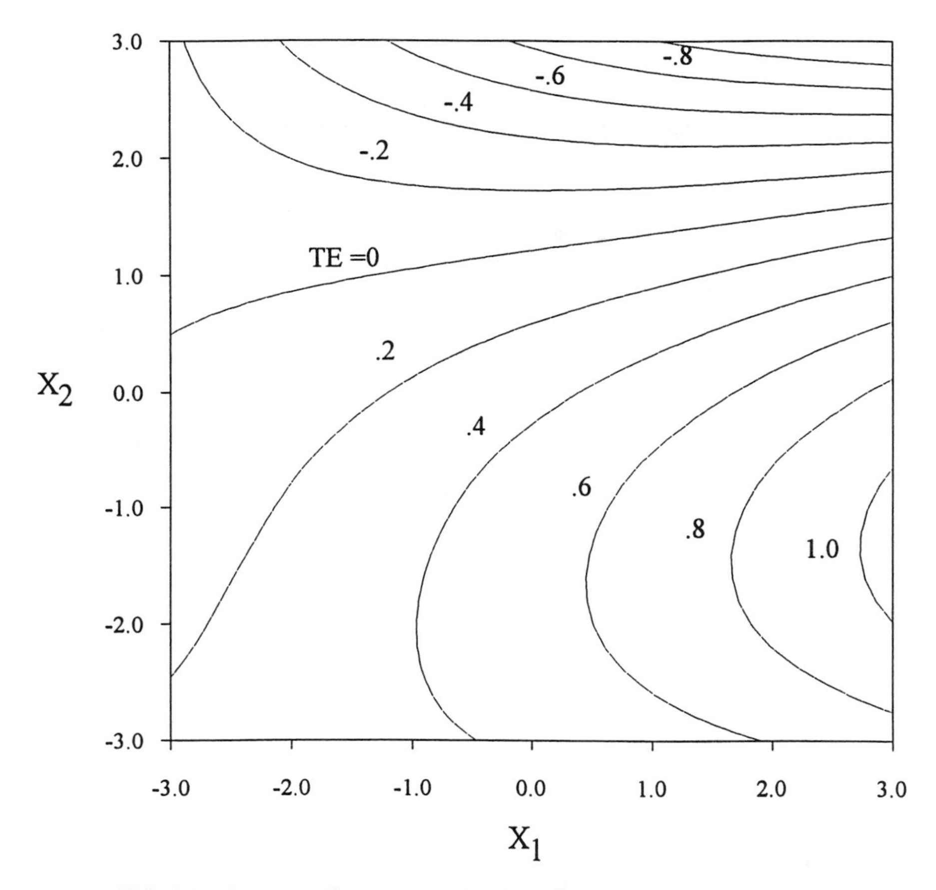

FIG. 8.4. Constant effect contours for the systematic portion of the total effect of x_2 on y_3.

and +1.0, it is seen that the total effect is in the range of 0.2 – 0.4, with varying x having a relatively weak impact on the total effect. However, for higher xs above +1.0, the total effect rapidly drops to zero and then becomes negative, approaching –0.4 for those with xs 2 *SD*s below the mean.

Thus far, the description of the total effect of x_2 on y_3 (i.e., Equation 9) has focused on the systematic portion, ignoring the random component. It is seen from Equation 9 that a description of the impact of the random residuals of the endogenous variables (i.e., the contribution of the terms in square brackets, called for short "impact") involves four variables, the two exogenous variables and two random residuals. To describe this impact, I created plots of constant contours of impact as a function of x_1 and x_2 for different combinations of ζ_1 and ζ_2. For example, Fig. 8.5 shows such a contour plot for ζ_1 and ζ_2 values 2 *SD*s below their means. Within a plus

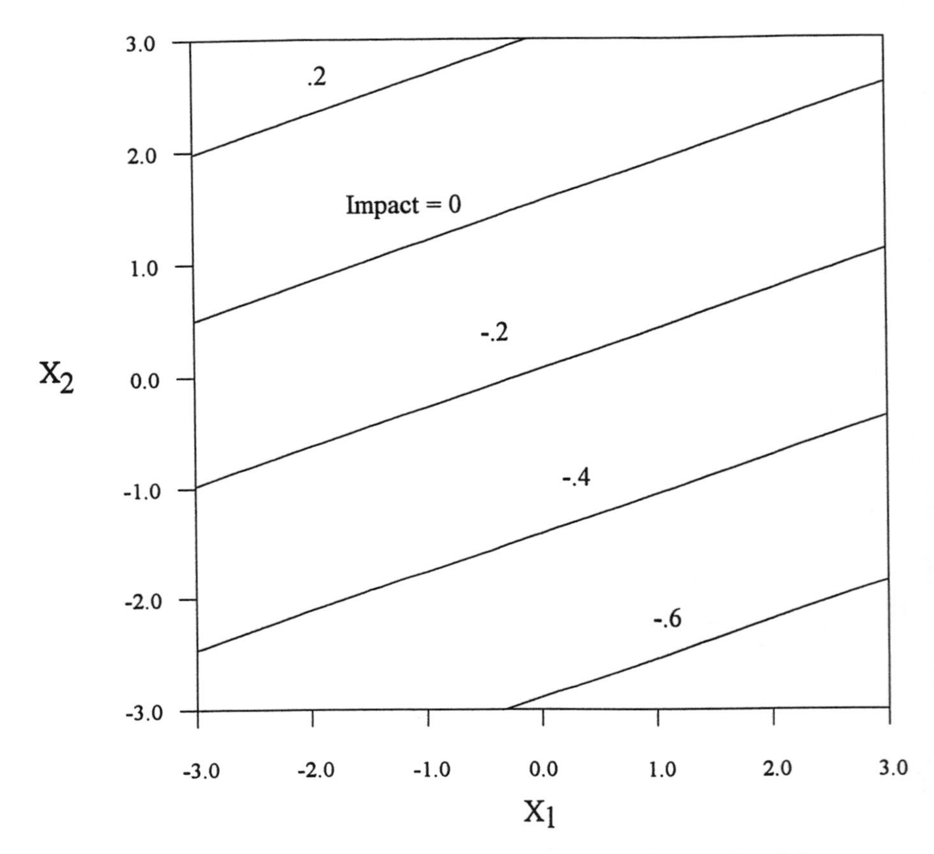

FIG. 8.5. Constant contours for the impact of the random portion of the total effect of x_2 on y_3 when the values of the random residuals for y_1 and y_2 are 2 *SD*s below the mean.

and minus 2 *SD* variation on the *x*s, the impact of the random errors is mostly negative, ranging up –0.5 in magnitude. The total range of impact over all similar plots with other ζ values was roughly represented by that shown in Fig. 8.5.

What is the importance of the random component of the total effect in Equation 9 relative to the systematic? Expressed differently, how serious is the possible degree of uncertainty introduced by the random component? Consider, for example, individuals having $x_1 = +1$ and $x_2 = -1$. From Fig. 8.4, the systematic portion of the total effect is approximately 0.6. For individuals having $\zeta_1 = \zeta_2 = 0$, the random component impact is zero, and for those with $\zeta_1 = \zeta_2 = -2$ *SD*s, the impact from Fig. 8.5 is approximately –0.4. Thus, for individuals with this specific *x* combination, the total effect would be either 0.6 or 0.2 depending on the ζ values, a quite large difference.

Of course, there would be many circumstances where the uncertainty would be less and there would be others where it is greater. It appears, though, that there is the potential for a discouraging degree of uncertainty in describing effects if one were to know the structural coefficients exactly.

The other total effects of *x*s on *y*s would be determined in a similar fashion, with associated indirect effects then found by subtraction. For the current example, there would be a total of 13 nonzero and nonredundant direct, indirect, and total effects, with all but 1 of those most likely involving a graphical representation. There would be various logical ways to combine some of these effects into a single graph. For example, the direct and total effects of one variable on another might be represented as two surfaces on a single 3-D plot with axes corresponding to the effect and the two exogenous variables. With this plot, it would be pointed out to the reader that the indirect effect is the vertical separation between the surfaces for the direct and total effects. Effective communication of results would be facilitated by judicious choice of graph format (e.g., 2-D, 3-D, or contour) to best highlight important features.

DISCUSSION

An interesting and limiting feature of effect decomposition for interactive/nonlinear models is the inherent uncertainty about indirect and total effects of exogenous variables on endogenous variables. As illustrated earlier (see, e.g., Equation 9), even if one knows the true values of all model coefficients, the effect of random residuals will still result in uncertainty about indirect and total effects. For the current example, the R^2 values of 0.64, 0.60, and 0.55 for the endogenous variables were respectable relative to values found in published research. Nevertheless, the degree of effect uncertainty associated with the residual terms was discouragingly large, resulting in some cases where an effect may vary randomly from weak to very strong for different individuals. The solution to this problem is obvious, but hardly simple. The degree of uncertainty can be reduced by improving specification of the structural equations to result in reduced error variability.

The procedure described here and the associated illustration have been limited to a recursive SEM of observed variables. A question for further study is whether the same procedure can be used for more general SEM models. At first glance, the matrix formulation for effect decomposition (e.g., Bollen, 1989) used here would also be appropriate for nonrecursive models, as it was in Bollen's presentation. However, it would seem that the required demonstration of stability of the system would be more problematic for interactive/nonlinear models. The single computation of eigenvalues of **B** (or, as implemented in LISREL, computation of the eigenvalues of **BB′**)

required in an additive/linear nonrecursive model would, it appears, transform to a requirement of determination of eigenvalues for each of many matrices corresponding to different combinations of the variables involved in the direct effect expressions. There may be more elegant solutions, but a "brute force" approach would simply compute the eigenvalues of **BB′** for each of a set of variable combinations roughly spanning the range of interest.

For a different type of extension, the effect decomposition procedure will, in principle, also work for recursive interactive/nonlinear models with latent variables. The basic logic of indirect effects resulting from the transmission of changes through a network of causal links is identical whether the structural links are between observed variables or between latent variables. In practice, of course, effect decomposition for interactive/nonlinear models with latent variable will have to wait on the availability of general and accessible estimation procedures.

SUMMARY

Most researchers in the behavioral and social sciences would accept the following two premises: (a) The complexity of human behavior demands the use of models that allow functional complexities such as interactions and nonlinearities, and (b) the logical implications of a hypothesized causal model can only be fully understood when indirect and total causal effects are considered and described. It is therefore surprising that so little attention has been given to effect decomposition for interactive/nonlinear SEM. A straightforward approach for such effect decomposition has been described and illustrated here. The procedure was limited to recursive interactive/nonlinear SEM with observed variables. It was argued that the same approach could be used for recursive latent-variable SEM once the associated estimation problems have been solved. It was also speculated that the approach may be appropriate for nonrecursive models.

REFERENCES

Aiken, L. S., & West, S. C. (1991). *Multiple regression: Testing and interpreting interactions.* Newbury Park, CA: Sage.

Bollen, K. A. (1987). Total, direct, and indirect effects in structural equation models. In C. C. Clogg (Ed.), *Sociological methodology 1987* (pp. 37–69). Washington, DC: American Sociological Association.

Bollen, K. A. (1989). *Structural equations with latent variables.* New York: Wiley.

Cohen, J., & Cohen, P. (1983). *Applied multiple regression/correlation analysis for the behavioral sciences* (2nd ed.). Hillsdale, NJ: Lawrence Erlbaum Associates.

Fox, J. (1985). Effect analysis in structural equation models II: Calculation of specific indirect effects. *Sociological Methods and Research, 14,* 81–95.

Hayduk, L. A. (1987). *Structural equation modeling with LISREL: Essentials and advances.* Baltimore, MD: Johns Hopkins University Press.

Heise, D. R. (1986). Estimating nonlinear models correcting for measurement error. *Sociological Methods and Research, 15,* 447–472.

Jaccard, J., Turrisi, R., & Wan, C. K. (1990). *Interaction effects in multiple regression.* Newbury Park, CA: Sage.

Jaccard, J., & Wan, C. (1995). Measurement error in the analysis of interaction effects between continuous predictors using multiple regression: Multiple indicator and structural equation approaches. *Psychological Bulletin, 117,* 348–357.

Jaccard, J., & Wan, C. (1996). *LISREL approaches to interaction effects in multiple regression* (Sage University Papers series on Quantitative Applications in Social Sciences). Beverly Hill, CA: Sage.

Jöreskog, K. G., & Sörbom, D. (1989). *LISREL 7: A guide to the program and applications* (2nd ed.). Chicago: SPSS Publications.

Jöreskog, K. G., & Yang, F. (1996). Nonlinear structural equation models: The Kenny–Judd model with interaction effects. In G. A. Marcoulides & R. E. Schumacker (Eds.), *Advanced structural equation modeling: Issues and techniques* (pp. 57–88). Mahwah, NJ: Lawrence Erlbaum Associates.

Kenny, D. A., & Judd, C. M. (1984). Estimating the non-linear and interactive effects of latent variables. *Psychological Bulletin, 96,* 201–210.

Lance, C. E. (1988). Residual centering, exploratory and confirmatory moderator analysis, and decomposition of effects in path models containing interactions. *Applied Psychological Measurement, 12,* 163–175.

Marsden, P. V. (1981). Conditional effects in regression models. In P. V. Marsden (Ed.), *Linear models in social research* (pp. 97–122). Beverly Hills, CA: Sage.

Ping, R. A., Jr. (1996a). Latent variable interaction and quadratic effect estimation: A two-step technique using structural equation analysis. *Psychological Bulletin, 119,* 166–175.

Ping, R. A., Jr. (1996b). Latent variable regression: A technique for estimating interaction and quadratic coefficients. *Multivariate Behavioral Research, 31,* 95–120.

Saunders, D. R. (1956). Moderator variables in prediction. *Educational and Psychological Measurement, 16,* 209–222.

Stolzenberg, R. M. (1980). The measurement and decomposition of causal effects in nonlinear and nonadditive models. In K. Schuessler (Ed.), *Sociological methodology 1980* (pp. 459–488). San Francisco: Jossey-Bass.

Tate, R. L. (1984). Limitations of centering for interactive models. *Sociological Methods & Research, 13,* 251–271.

9

Estimating Nonlinear Effects Using a Structured Means Intercept Approach

Benoît Laplante
Institut National de la Recherche Scientifique

Stéphane Sabourin
Université Laval

Louis-Georges Cournoyer
John Wright
Université de Montréal

Since the pioneer work of Kenny and Judd (1984), two problems have precluded the wide use of structural equation models (SEMs) that include nonlinear relations among latent continuous variables.

The first problem is that most, if not all, of the methods that have been proposed to specify this type of nonlinear relations use products of observed variables as multiple indicators of the product term of the latent independent variables involved in the nonlinear relation. This strategy typically leads to computing a large number of product variables whose measurement errors are correlated with the measurement errors of the original variables, and to explicitly specifying these covariances in the model. Hayduk (1987) provided a classical example of this; Jaccard and Wan (1995, 1996), Ping (1995, 1996), as well as Jöreskog and Yang (1996) are more recent examples. Explicitly specifying these covariances and all the linear and nonlinear equalities they induce among the parameters in the structural model is a cumbersome and error-prone process that few researchers seem ready to undertake routinely.

The second problem is more fundamental and stems from the nature of nonlinear relations among continuous variables. Unlike linear relations, nonlinear relations, whether they are considered in correlation, regression, or SEMs, are not scale insensitive.

In this chapter, we deal with these two problems, and we propose a solution for each of them. Assuming that the latent exogenous variables are multivariate normal, and that they are measured without bias, we show how to specify interactions and curvilinear relations without resorting to products of observed variables, and without leaving the framework of structural equation modeling. This is done with as few nonlinear contraints as possible and one "phantom" variable for each nonlinear relation. Assuming that the measurement errors of the latent endogenous variables are not correlated with the measurement errors and the prediction error of the latent endogenous variable, we show that it is possible to estimate the intercept of a structural equation with ordinary multiple regression, and to use this information to fully specify and estimate the structural model. Combined together, these two solutions provide a complete procedure for the specification and estimation of SEMs that include nonlinear relations. Finally, we assess the ability of this procedure to correctly estimate such models through a Monte Carlo study.

THE SCALE BIAS PROBLEM IN THE ESTIMATION OF INTERACTIONS

What is known in the literature as "the scale bias problem in the estimation of interactions" is a manifold problem. One aspect of it is a consequence of a quite simple property of products: The product of two negative quantities gives the same result as the product of two positive quantities of same absolute values, for instance $(-2)(-6) = (2)(6) = 12$. This property does not create any problem in the usual cases of correlation or regression; it does, however, when one uses the product of two exogenous variables as a new exogenous variable. In such a case, the values of this new variable for each individual in the sample will depend greatly on the coding system. More specifically, if both original variables have only positive values, people having low values on both will get low values for the product variable, whereas if both original variables have negative values, people having low scores on both will get the same values as people having high scores on both. In other words, the product of two variables having negative scores is an arithmetic operation that does not retain the order of the original values. One consequence of this is to create an artificial interaction: For instance, in a case where the effects of two exogenous variables on one endogenous variable are truly linear, testing a model that includes an interaction by using a product term computed from deviation or standardized scores would show evidence of an interaction with estimates that would make the effect of each exogenous variable negative for low values and positive for high values (Tate, 1984). Another aspect of the problem, raised by Allison (1977), is that values of zero in any of the exogenous variables involved in an interaction eliminate the effects of both variables for any combination including such a value. A third aspect, studied by Bohrnstedt and Marwell (1978), is that the reliability of a product of variables depends on the correlation of the two variables, but also on the ratio of the means of these variables to their respective variances: A change in the magnitude of the scale of the variables will not affect the reliability of their product, but adding a constant to them will. Still another part of the problem is that the mean of an endogenous variable that is dependent on an interaction among two exogenous variables will not generally be zero, even if the two exogenous variables have zero means (Jöreskog & Yang, 1996).

Although some recent contributions on nonlinear relations in SEMs do not address these questions (Ping 1995, 1996), various solutions have been proposed to overcome some of these difficulties. One element that is part of many of these solutions is to use SEMs with structured means (Jaccard & Wan, 1996; Jöreskog & Yang, 1996). However, because in such models, the intercepts of the structural equations are not readily identified, the model cannot be estimated without adding further information to it. Assigning zero to the intercept of the structural equations, as suggested by Jaccard and Wan, only makes the problem worse, because, as in ordinary regression, the means of the latent variables and the parameters of the structural equations are algebraically related to the intercept.

Jöreskog and Yang (1996) have provided a thorough treatment of the example used by Kenny and Judd (1984). They made it clear that, as a rule, the intercept of a structural equation that includes a nonlinear relation has to be estimated even when the latent exogenous variables involved in this relation both have zero means. They showed that, under certain assumptions, the SEM derived from this example can be fully identified and estimated, including the intercept of the structural equation and the systematic measurement error components of the observed variables, by using only one product of observed variables. However, this strategy has limitations of its own. Most of their discussion and examples assumed that the means of the latent exogenous variables involved in the interaction are zero: Such an assumption is unlikely to be true in most cases, and we discussed earlier why centering should be avoided in the specification of nonlinear relations. Finally, as they pointed out themselves, the constraints needed to specify their models are quite complex. An example that includes only two exogenous latent variables, each of which has zero mean, where the random measurement errors of the observed endogenous variables are mutually uncorrelated, and where the endogenous variable has only one indicator requires five nonlinear constraints. An example with two latent exogenous variables, each measured with two observed variables, and one latent endogenous variable, measured with three observed variables, needs no less than 7 linear and 22 nonlinear constraints in order to be completely specified.

Another solution, proposed by Bollen (1995), uses products of observed variables, avoids the specification of the covariances between their measurement errors, and provides estimates of the intercepts of the structural equations. However, by relying on a two-stage least squares estimator developed by the same author (Bollen, 1996), this solution forces its user to leave the conventional framework of structural equation modeling: It requires a rather lengthy process of step-by-step estimation, bypasses the estimation of the measurement portion of the SEM, and makes difficult the computation of most of the statistics commonly associated with the global fit of SEMs (also see Bollen & Paxton, chap. 6; Li & Harmer, chap. 7).

The solution we propose does not assume that the exogenous latent variables have zero means, which avoids the so-called centering problem. It includes a procedure to estimate the intercept of the structural equation. It does not use products of observed variables, and therefore limits the number of contraints that have to be imposed on the parameters, even when each of the latent variables are measured with many indicators. It does not leave the framework of structural equation modeling, and thus can be used with any SEM software that handles nonlinear constraints.

AN EXAMPLE INCLUDING AN INTERACTION

The example we use throughout the remainder of this chapter is rather simple. A latent endogenous variable η_1 is thought to be dependent on three latent exogenous variables ξ_1, ξ_2, and ξ_3. The effect of ξ_3 on η_1 is linear whereas the relation between ξ_1, ξ_2, and η_1 is an interaction; that is, $\gamma_{\eta_1\xi_1} = f(\xi_2)$ and $\gamma_{\eta_1\xi_2} = g(\xi_1)$. Each of the latent exogenous variables is measured with three observed variables, whereas the latent endogenous variable is measured with five different indicators. We allow the latent exogenous variables to be mutually correlated. We assume that the measurement errors are normal random variates. The measurement errors are not correlated. However, this is not an assumption, but a simple convenience: Our specification procedure does not require the measurement errors to be mutually uncorrelated. We use a structured means model; in other words, we do not express the values of the observed variables as deviations from their means, and we do not impose zero means to the latent variables. However, in order to build a model that can be identified, we assume that the indicators measure the latent variables without bias. Using Jöreskog's notation, the structural portion of this model is described by the following equation:

$$\eta_1 = \alpha_1 + \gamma_{11}\xi_1 + \gamma_{12}\xi_2 + \gamma_{13}\xi_3 + \gamma_{14}\xi_4^* + \zeta_1, \quad (1)$$

where $\xi_4^* = \xi_1 \times \xi_2$. The effect of ξ_1 on η_1 is then $\gamma_{\eta_1\xi_1} = \gamma_{11} + \gamma_{14}\xi_2$, and the effect of ξ_2 on η_1 is $\gamma_{\eta_1\xi_2} = \gamma_{12} + \gamma_{14}\xi_1$.

The measurement portion of the model is described by the two following equations:

$$\mathbf{y} = \Lambda_y \eta + \varepsilon \tag{2}$$

and

$$\mathbf{x} = \Lambda_x \xi + \delta. \tag{3}$$

As stated previously, this example is a special case of a SEM with structured means where the latent variables are measured without systematic error. In such a model, the means of the observed variables are given by

$$\mu_x = \Lambda_x \kappa \tag{4}$$

and

$$\mu_y = \Lambda_y \eta(\alpha + \Gamma\kappa), \tag{5}$$

where κ is the mean vector of the latent exogenous variables.

As in all SEMs, this model will be identified only if the latent variables are scaled in some way. This can easily be done by imposing a unit value to one element in each column of Λ_y and Λ_x, as routinely done with more common models. The variance of each latent variable will then be equal to the variance of the observed variable used to scale it minus the portion of the variance of this observed variable that comes from random measurement error. However, and this is an important feature, because we assume that the latent variables are measured without bias, this procedure will also assign the means of these observed variables to the latent variables they are scaling.

THE MISUSE OF PRODUCTS OF OBSERVED VARIABLES

If we were to use the method introduced by Kenny and Judd (1984) to specify this model, we would need to compute the products between each of the indicators of the first latent variable and each of the indicators of the second latent variable, and to use these product variables as indicators of the product of the two latent variables involved in the interaction. In our example, where ξ_1 and ξ_2 each have three indicators, X_1 to X_3 and X_4 to X_6, ξ_4^* would have nine indicators, say X_7 to X_{15}, that would be equal to X_1X_4, X_1X_2, X_1X_3, X_2X_4, and so forth. The proper estimation of SEMs requires that equalities and known algebraic relations between the parameters be explicitly specified. It is quite obvious that many such algebraic relations exist in the system we just described: The variance of ξ_4^* is a function of the variances of ξ_1 and ξ_2, and of their covariance, and the variances of X_7 to X_{15} are functions of the variance of ξ_4^*, of the variances of ξ_1 and ξ_2, and of the variance of the measurement errors of their indicators. Taking means into consideration would force us to include further constraints.

To write down and program all of these functions is a quite formidable task, and involves a lot of cumbersome and error-prone algebraic manipulations. The amount of work needed to do so has probably discouraged many reasearchers to include interactions in SEMs. However, most of these specifications are not needed. The method introduced by Kenny and Judd (1984) relies on the misconception that SEMs are barely sophisticated forms of multiple regression, and that the "natural" way to specify nonlinear relations in a SEM is to use product terms of observed variables, as it is customary to do in regression analysis. However, structural equation modeling is not regression analysis: It models relations between the parameters. To be specific, one of the basic features of structural equation modeling is to include the variances, covariances, and, under certain conditions, the means of the latent variables as parameters of the model, whereas these quantities are merely data or by-products in regression analysis. If one needs to include the product, or any other function, of two latent variables as a term in a structural equation, one does not need indicators of this new term: One just needs to specify correctly the parameters pertaining to it as functions of the other relevant parameters of the model. In other words, the modelization features of structural modeling may be used either to take into account explicitly the covariances between the measurement errors of the latent predictors and those of the indicators of the product of the latent predictors, as Kenny and Judd (1984) suggested, or, as we propose, to eliminate the products of observed variables and all their induced covariances. The latter solution is simpler in every respect, and we favor it for this reason. Implementing this solution amounts to using a structural model that includes means, and to specify correctly the means, variances, and covariances of the latent variables that are functions of the means, variances, and covariances of some other latent variables. We now develop these topics.

THE SPECIFICATION OF THE INTERACTION EXAMPLE

The measurement portion of our SEM is quite conventional. Assuming that the latent variables are measured without systematic error ensures that all the parameters involved in it are identified. As explained earlier, this assumption also implies that the means of the latent variables will be the means of the observed variables that are used to scale them. In other words, all the means of the latent exogenous variables, that is all the elements of κ but κ_4^*, the mean of $\xi_1\xi_2$, are identified. This parameter is the first that we must specify as a function of other parameters of the model. This is a rather simple task: Knowing that the mean of the product of two variables x and y is simply

$$E(xy) = E(x)E(y) + C(x,y), \tag{6}$$

where E is the expectancy operator and C, a covariance, we may write

$$\kappa_4^* = \kappa_1\kappa_2 + \phi_{12}. \tag{7}$$

The correct specification of the variances and covariances that are functions of other parameters of the model is a more demanding matter. As stated earlier, we wish to allow all the original exogenous latent variables to be mutually correlated. We also know that the latent product term is a nonlinear function of some of these variables. This makes it imperative to specify correctly not only its variance, but also its covariance with all the other exogenous latent variables. Therefore, in our example, four more parameters need to be properly expressed as functions of other parameters: the mean and the variance of the product of ξ_1 and ξ_2 (respectively κ_4^* and ϕ_{44}^*), and the covariances between this product and ξ_1, ξ_2, and ξ_3 (ϕ_{41}^*, ϕ_{42}^*, and ϕ_{43}^*).

From Goldman (1960), we know that the variance of the product of two variables x and y can be expressed as

$$\begin{aligned} V(xy) = {} & E^2(x)V(y) + E^2(y)V(x) + E[(\Delta x)^2(\Delta y)^2] \\ & + 2E(x)E[(\Delta x)(\Delta y)^2] + 2E(y)E[(\Delta x)^2(\Delta y)] \\ & + 2E(x)E(y)C(x,y) - C^2(x,y), \end{aligned} \tag{8}$$

where V denotes a variance and $\Delta x = x - E(x)$.

Some of the terms appearing in Equation 8, those involving only the expected values, variances, or covariance of the original variables, can readily be translated into parameters of a SEM. However many others cannot: There is no straightforward transformation that can express the expected value of the product of the squares of the deviations from the means of the original variables (the third term of the equation) as a combination of their means, variances, and covariance. The problem is the same with the fourth and fifth terms. Unless one is willing to use a reformulation of the mathematics of structural equation modeling that would include various combinations of the higher moments of the original variables as parameters, there is no other solution than to impose some simplifying assumptions to the form of the bivariate distribution of x and y. Bohrnstedt and Goldberger (1969) have shown that, assuming bivariate normality, Equation 8 reduces to

$$\begin{aligned} V(xy) = {} & E^2(x)V(y) + E^2(y)V(x) + 2E(x)E(y)C(x,y) \\ & + V(x)V(y) + C^2(x,y). \end{aligned} \tag{9}$$

Each term of this equation matches a parameter of the SEM we are using as an example. This enables us to write

$$\phi^*_{44} = \kappa_1^2\phi_{22} + \kappa_2^2\phi_{11} + 2\kappa_1\kappa_2\phi_{21} + \phi_{11}\phi_{22} + \phi_{21}^2. \quad (10)$$

Bohrnstedt and Goldberger (1969) stated that the covariance between the product of two variables x and y, and a third variable z is given by

$$C(xy,z) = E(x)C(y,z) + E(y)C(x,z) + E[(\Delta x)(\Delta y)(\Delta z)], \quad (11)$$

which, under bivariate normality, reduces to

$$C(xy,z) = E(x)C(y,z) + E(y)C(x,z), \quad (12)$$

and enables us to write

$$\phi^*_{43} = \kappa_1\phi_{31} + \kappa_2\phi_{32}. \quad (13)$$

We do not know of any published results on the covariance between a variable and its product with another variable, but, using the same kind of approach as Bohrnstedt and Goldberger (1969) did, one finds that (see Appendix A for a proof)

$$C(x,xy) = E(x)C(x,y) + E(y)V(x) + E[(\Delta x)^2(\Delta y)], \quad (14)$$

which, again assuming bivariate normality, reduces to

$$C(x,xy) = E(x)C(x,y) + E(y)V(x) \quad (15)$$

and allows us to write

$$\phi^*_{41} = \kappa_1\phi_{21} + \kappa_2\phi_{11} \quad (16)$$

and

$$\phi^*_{42} = \kappa_2\phi_{21} + \kappa_1\phi_{22}. \quad (17)$$

Equations 7, 10, 13, 16, and 17 express the only (five) special constraints that have to be imposed on the parameters in order to specify, and correctly estimate the model described in Equation 1. Equations 6, 9, 12, and 15 should be the only tools needed to specify interactions between two latent exogenous variables and one latent endogenous variable in most SEMs.

THE SPECIFICATION OF CURVILINEAR EFFECTS AMONG THE LATENT VARIABLES

Much of our discussion of interaction and of the problems of its specification in SEMs, as well as our solution to these, can be applied to curvilinear relations. A curvilinear relation is a relation whose sign and magnitude depends on some powers of the values of the independent variable, that is, $\gamma_{\eta_1\xi_1} = h(\xi_1, \xi_1^2, \ldots \xi_1^n)$. In its simplest form, the relation between the effect and the values of the independent variable is assumed to be linear, and to involve only the first power of the variable. Thus $\gamma_{\eta_1\xi_1} = \gamma_{11} + \gamma_{13}\xi_1$. A model somewhat similar to Equation 1, but that includes only two independent variables and a curvilinear relationship between ξ_1 and η_1 can be described by

$$\eta_1 = \alpha_1 + \gamma_{11}\xi_1 + \gamma_{12}\xi_2 + \gamma_{13}\xi_3^* + \zeta_1, \tag{18}$$

where $\xi_3^* = \xi_1^2$.

Four parameters must be expressed as functions of other parameters of the model in order to specify it correctly. These are κ_3^*, the mean of the square of the latent independent variable involved in the curvilinear relation, ϕ_{33}^*, its variance, ϕ_{31}^*, its covariance with its square root ξ_1, and finally ϕ_{32}^*, its covariance with the other exogenous latent variable.

Using Equation 6, we can write

$$E(x^2) = E^2(x) + V(x) \tag{19}$$

and thus

$$\kappa_3^* = \kappa_1^2 + \phi_{11}. \tag{20}$$

Using Equation 9, we can write

$$V(x^2) = 4E^2(x) + 2V^2(x) \tag{21}$$

and

$$\phi_{33}^* = 4\kappa_1^2\phi_{11} + 2\phi_{11}^2. \tag{22}$$

Equation 12 enables us to write

$$C(x^2,x) = 2E(x)V(x) \tag{23}$$

and

$$\phi^*_{31} = 2\kappa_1\phi_1. \tag{24}$$

Finally, using Equation 15 leads to

$$C(x^2,z) = 2E(x)C(x,z) \tag{25}$$

and

$$\phi^*_{32} = 2\kappa_1\phi_{21}. \tag{26}$$

Equations 20, 22, 24, and 26 express the only (four) constraints that have to be imposed on the parameters of the model to specify the curvilinear relation included in Equation 18. Equations 19, 21, 23, and 25 should be the only tools needed to specify a second-degree curvilinear relation in most SEMs. Deducting the constraints needed to specify curvilinear relations of higher degrees is straightforward.

THE INTERCEPT OF THE STRUCTURAL EQUATION MODEL

SEMs are an extension of covariance structure analysis. Their data are aggregate statistics, such as means, variances, and covariances, and not individual-level information. It is possible to recast, at least in part, SEMs into the framework of individual data analysis: Bentler (1993) and Bollen (1995, 1996) provided examples of such attempts. When using individual-level data, intercept estimation is not a problem. However, the intercepts of the structural equations are not identified in the conventional framework of structural equation modeling, and, therefore, SEMs that include them are not identified, and cannot be readily estimated.

One way to overcome this problem is to impose an arbitrary value, such as zero, to the intercept, as suggested in Jaccard and Wan (1996). However, this solution provides biased estimates of the other parameters in the equation unless the true value of the intercept is zero, which is seldom the case. A simple way to make this clear is to take the expectancies of Equation 1:

$$E(\eta_1) = E(\alpha) + \gamma_{12}E(\xi_1) + \gamma_{12}E(\xi_2) + \gamma_{13}E(\xi_3) + \gamma_{14}E(\xi^*_4). \tag{27}$$

Most of these quantities are parameters of a SEM. We can therefore rewrite Equation 27 as follows:

$$E(\eta_1) = \alpha + \gamma_{11}\kappa_1 + \gamma_{12}\kappa_2 + \gamma_{13}\kappa_3 + \gamma_{14}\kappa^*_4. \tag{28}$$

When applied to sample data, this equation becomes:

$$\bar{Y}_1 = \alpha + \gamma_{11}\bar{X}_1 + \gamma_{12}\bar{X}_4 + \gamma_{13}\bar{X}_7 + \gamma_{14}\overline{X_1X_4}. \tag{29}$$

From there, it is easy to see that, in SEMs as in multiple regression, the intercept of a structural equation is equal to the mean of the dependent variable minus the average effect of each independent variable, that is

$$\alpha = \bar{Y}_1 - \gamma_{11}\bar{X}_1 - \gamma_{12}\bar{X}_4 - \gamma_{13}\bar{X}_7 - \gamma_{14}\overline{X_1X_4}. \tag{30}$$

This constraint on the intercept should make clear that assigning any given value to it while keeping the means of the observed variables unchanged will provide a different set of estimates of the structural parameters for each assigned value. In other words, the line, in the linear case, or curve, in the nonlinear case, of the true structural relation has a definite origin, and assigning an arbitrary value to this origin does not simply slide this line or curve along the ordinate axis without changing its slope or curvature; it affects its slope or curvature as well.

In their work on the Kenny–Judd model, Jöreskog and Yang (1996) made clear that, as a rule, the intercept of a structural equation that includes a nonlinear relation is not zero even when the latent exogenous variables involved in this relation both have zero means. They showed that it is possible to fully identify and estimate the example used by Kenny and Judd, including the intercept of the structural equation and the systematic measurement errors of the observed exogenous variables, by using only one product of observed exogenous variables. However, as we pointed out earlier, most of their discussion and examples assumed that the means of the latent exogenous variables involved in the interaction are zero, which is unlikely to be true in most cases and should be avoided in the specification of nonlinear relations. Finally, even with this simplifying assumption, the constraints needed to specify the models are quite complex.

Another way to overcome the unidentification of the intercept is to estimate the structural equations directly from individual data, as proposed by Bollen (1995, 1996). This procedure provides unbiased estimates of the structural parameters, including the intercepts, under a wide variety of conditions. However, it has a few drawbacks. It estimates directly the structural parameters through regression, and thus bypasses the estimation of the measurement portion of the SEM; therefore, it provides no information about the factor analytic component of the SEM. It requires that each structural equation be expressed as a function of observed variables, which demands some error-prone algebraic manipulations. It uses instrumental variables to estimate the structural parameters, which requires the researcher to infer, from the sample data, which observed and latent variables are correlated, and which are not, and to select from the latter those that can be used as instruments in the estimation. The structural parameters are estimated in multiple steps with regression software, which deprives the user from most of the fit and assessment statistics that are routinely computed by overall estimation procedures.

A simpler way to overcome the unidentification of the intercept of a structural equation is to estimate this parameter before estimating the structural model itself, and to assign the resulting estimate to the intercept when estimating the other parameters and the fit of the whole model. What follows shows that, with one assumption, it is possible and quite easy to obtain such estimates for models similar to the one we are dealing with.

Our model supposes that $Y_1 = \eta_1 + \varepsilon_1$. From this and Equation 1, we can write

$$Y_1 = (\alpha_1 + \gamma_{11}\xi_1 + \gamma_{12}\xi_2 + \gamma_{13}\xi_3 + \gamma_{14}\xi_1\xi_2 + \zeta_1) + \varepsilon_1. \tag{31}$$

If we assume that, in our model, the δs are not correlated with ζ and the εs, the elements of Γ can be interpreted as disattenuated regression coefficients. In other words, we can rewrite Equation 31 as

$$\begin{aligned} Y_1 = {} & \alpha_1\rho_0 + (b_{1,1}X_1)\rho_1 + (b_{1,4}X_4)\rho_4 + (b_{1,7}X_7)\rho_7 \\ & + (b_{1,10}X_1X_4)\rho_{10} + \zeta + \varepsilon_1, \end{aligned} \tag{32}$$

where each latent exogenous variable is replaced by its scaling observed variable, and the product of the two exogenous latent variables involved in the nonlinear relation is replaced by the product of their scaling variables. In Equation 32, a "b" is therefore a regression coefficient, each ρ is the reliability of an observed exogenous variable relative to its corresponding latent exogenous variable, and $\zeta + \varepsilon_1$ is equal to the residual of the regression equation. Because we assume that the latent exogenous variables are measured with error, the regression coefficients of Equation 32 are not unbiased estimates of the corresponding structural coefficients. However, because the intercept, in both the structural equation and the regression equation, is a parameter related to the same constant, its reliability is 1, and the intercept of the regression equation is an unbiased estimate of the intercept of the structural equation. In other words, assuming that, in the true SEM, the measurement errors of the observed exogenous variables are not correlated with the measurement errors of the observed endogenous variables and the residual of the structural equation, the intercept of the regression of the scaling variable of an endogenous latent variable on the scaling variables of a set of latent exogenous variables is an estimate of the otherwise unidentified intercept of the corresponding structural equation. This result is not limited to structural equations involving nonlinear relations: It applies to any structural equation that includes an intercept. Assigning the value of this estimate to the intercept of the structural equation makes the model fully identified even though we do not use any product of observed variables in the model itself, and allows one to estimate the other parameters as well as the global fit with conventional SEM estimation procedures. Standard errors and other test statistics should be unbiased as long as they are computed with the correct number of degrees of freeedom, which should be that of a conventional model minus the number of previously estimated intercepts.

A MONTE CARLO STUDY OF THE INTERACTION EXAMPLE

To assess the ability of our procedure to correctly estimate the parameters of a structural model involving a nonlinear relationship among the latent variables, we tested it on artificial data of known structure. The structure we chose for this study is based on the example we used in our discussion of interaction. Three exogenous latent variables are each measured with three observed variables, and one endogenous latent variable is measured with five observed variables. The product needed to specify the interaction in the structural equation has no observed variable of its own. The theoretical population covariance matrix was computed from arbitrary values for each parameter. One hundred artificial samples of 2,000 cases were generated using the Cholesky decomposition of the theoretical covariance matrix and 2,000 random normal vectors of 14 elements each. The artificial covariance matrices and mean vectors used as data for the analyses were computed from the artificial samples. The estimations were conducted using PROC REG and PROC CALIS, two components of the SAS/STAT software; the code needed to estimate the structural model is given in Appendix B. The results of this Monte Carlo study are reported in Table 9.1.

The first column of Table 9.1 lists the name of each parameter of the model. The second column contains the known population values of the parameters. The third column contains the mean of the estimates computed from the artificial data; the fourth column contains the means of their standard errors.

The estimates of the measurement portion of the model are close to the theoretical values, as well as the estimate of the effect of the third latent exogenous variable on the latent endogenous variable. The variance of the latent product is slightly underestimated, as well as its covariances with the other latent exogenous variables. The intercept of the structural equation is also slightly underestimated. In the theoretical model, the effect of ξ_1 on η_1 is $\gamma_{\eta_1\xi_1} = 1.000 + .100\xi_2$, and the effect of ξ_2 on η_1 is $\gamma_{\eta_1\xi_2} = 1.000 + .100\xi_1$. The estimates from artificial data are $\gamma_{\eta_1\xi_1} = 1.248 + .058\xi_2$ and $\gamma_{\eta_1\xi_2} = 1.207 + .058\xi_1$, and therefore the interaction is underestimated.

TABLE 9.1
Monte Carlo Study of a Structural Equation Model Including an Interaction Among Latent Continuous Variables and Estimated Without Product of Observed Variables and With Prior Estimation of the Intercept of the Structural Equation

Parameters	*Theoretical Values*	*Means of Estimates*	*Means of Standard Errors*
κ_1	5.000	5.011	.043
κ_2	6.000	6.015	.052
κ_3	7.000	7.012	.060
κ_4^*	31.014[a]	31.122[b]	.417
ϕ_{11}	3.380	3.310	.112
ϕ_{22}	4.868	4.790	.163
ϕ_{33}	6.625	6.481	.223
ϕ_{44}^*	321.702[a]	316.165[b]	11.634
ϕ_{21}	1.014	.983	.098
ϕ_{31}	1.420	1.404	.115
ϕ_{32}	1.136	1.114	.136
ϕ_{41}^*	25.352[a]	24.837[b]	.972
ϕ_{42}^*	30.422[a]	29.918[b]	1.171
ϕ_{43}^*	14.197[a]	14.027[b]	1.100
γ_{11}	1.000	1.248	.140
γ_{12}	1.000	1.207	.109
γ_{13}	1.000	1.018	.073
γ_{14}	.100	.058	.024
α_1	50.000	48.697	1.463
ψ_1	50.000	49.440	1.769
λ_{y11}	1.000	[1.000]	n.a.
λ_{y21}	.900	.900	.002
λ_{y31}	.800	.800	.002
λ_{y41}	1.100	1.100	.002
λ_{y51}	1.200	1.200	.003
θ_{e1}	23.357	23.106	.886
θ_{e2}	13.355	13.191	.554
θ_{e3}	19.932	19.682	.717
θ_{e4}	19.950	19.752	.828
θ_{e5}	33.635	33.406	1.279
λ_{x11}	1.000	[1.000]	n.a.
λ_{x21}	1.100	1.101	.005
λ_{x31}	.900	.901	.005
λ_{x42}	1.000	[1.000]	n.a.
λ_{x52}	1.200	1.199	.005
λ_{x62}	1.100	1.100	.006
λ_{x73}	1.000	[1.000]	n.a.
λ_{x83}	.900	.901	.004
λ_{x93}	.800	.801	.005
$\theta_{\delta 1}$	.376	.370	.027
$\theta_{\delta 2}$	.714	.696	.037
$\theta_{\delta 3}$	1.352	1.307	.047

(Continued)

TABLE 9.1
(Continued)

Parameters	*Theoretical Values*	*Means of Estimates*	*Means of Standard Errors*
$\theta_{\delta 4}$	.541	.535	.039
$\theta_{\delta 5}$	1.501	1.460	.070
$\theta_{\delta 6}$	2.366	2.309	.087
$\theta_{\delta 7}$	.736	.750	.063
$\theta_{\delta 8}$	1.399	1.355	.065
$\theta_{\delta 9}$	2.650	2.540	.092

Note. Coefficients between square brackets were fixed to the value reported.
[a]Computed as a function of other parameters.
[b]Estimated as a function of other parameters.

LIMITATIONS AND DISCUSSION

We have shown that it is possible to specify SEMs involving nonlinear relations, such as interactions or curvilinear relations among one or several latent exogenous variables and one endogenous latent variable, without resorting to products of observed variables, even when each of the latent variables is measured with several indicators, and without leaving the conventional framework of structural equation modeling. This solution requires the use of a structured means model and two special assumptions: that the latent variables involved in the nonlinear relation be measured without systematic error, and that the latent exogenous variables involved in the latent product term be bivariate normal (the latent product term itself, of course, is not normal and not assumed to be normal).

We have also shown that, although the intercept of a structural equation that is part of a SEM with structured means is not readily identified, it is possible to estimate this intercept through regression, assuming that the measurement errors of the observed exogenous variables are not correlated with the measurement errors of the observed endogenous variables and the residual of the structural equation.

Finally, we have reported results of a Monte Carlo study that assesses the ability of the combined use of these two strategies to specify and estimate a SEM involving an interaction. Our results showed that the procedure correctly estimated the parameters of the measurement portion of the model as well as the linear effect of the third latent exogenous variable, but underestimated the coefficients describing the interaction.

There are no reasons to believe that the underestimation of the interaction is caused by the way we specified the model. Using products of observed variables and explicitly specifying in the model the covariances and equalities they induce would not reduce this bias: Jöreskog and Yang (1996) reported difficulties similar to ours, even with the use of an estimation procedure that explicitly takes non-normality into account.

As is usually the case in work on nonlinear relations in SEM, the main limitations of our procedure are its assumptions. Our procedure requires that the latent variables be measured without systematic errors, and that the random measurement errors of the observed endogenous variables as well as the prediction error of the latent endogenous variable not be correlated with the measurement errors of the observed exogenous variables. At this stage of our work, we do not see how to relax these assumptions.

Our procedure also requires that the latent exogenous variables be multivariate normal. We stress that this assumption is an assumption of the specification, not an assumption of the estimation method we used. It is necessary to assume multivariate normality of the latent exogenous variables to express the variance of their product and the covariance of this product with the other latent variables as functions of the parameters of a SEM. This assumption is common to all procedures that specify the variance of the latent product term using the Bohrnstedt and Goldberger (1969) formula. Because it is an assumption of the specification of the parameters of the model, it cannot be relaxed by the use of estimation procedures such as those based on Browne (1984), which are designed to obtain tests of fit and standard errors when the distributional assumption of other estimation methods are not satisfied.

Another limitation of our procedure is that it is limited to nonlinear relations among exogenous and endogenous variables. However, preliminary work indicates that it should be generalizable to nonlinear relations involving only endogenous variables.

ACKNOWLEDGMENTS

This study has been supported in part by grants from the Fonds FCAR and the Social Sciences and Humanities Research Council of Canada. An earlier version of this chapter was presented at the meeting of the Association canadienne des sociologues et anthropologues de langue française, Montreal, 1994. Special thanks to Denise Desrosiers who performed a substantial part of the work involved in the Monte Carlo study.

REFERENCES

Allison, P. D. (1977). Testing for interaction in multiple regression. *American Journal of Sociology, 83,* 144–153.

Bentler, P. M. (1993). Structural equation models as nonlinear regression models. In K. Haagen, D. Bartholomew, & M. Deistler (Eds.), *Statistical modeling and latent variables* (pp. 51–64). Amsterdam: Elsevier.

Bohrnstedt, G. W., & Goldberger, A. S. (1969). On the exact covariance of products of random variables. *Journal of the American Statistical Association, 64,* 1439–1442.

Bohrnstedt, G. W., & Marwell, G. (1978). The reliability of products of two random variables. In K. F. Schussler (Ed.), *Sociological methodology 1978* (pp. 254–273). San Francisco: Jossey-Bass.

Bollen, K. A. (1995). Structural equation models that are nonlinear in latent variables: A least-squares estimator. In P. Marsden (Ed.), *Sociological methodology 1995* (pp. 223–251). Cambridge, MA: Blackwell.

Bollen, K. A. (1996). An alternative two stage least squares (2SLS) estimator for latent variables equations. *Psychometrika, 61,* 109–121.

Browne, M. W. (1984). Asymptotically distribution-free methods for the analysis of covariance structures. *British Journal of Mathematical and Statistical Psychology, 37,* 62–83.

Goldman, L. A. (1960). On the exact variance of products. *Journal of the American Statistical Association, 55,* 708–713.

Hayduk, L. A. (1987). *Structural equation modeling with LISREL: Essentials and advances.* Baltimore: Johns Hopkins University Press.

Hayduk, L. A. (1996). *LISREL issues, debates, and strategies.* Baltimore: Johns Hopkins University Press.

Heise, D. R. (1986). Estimating nonlinear models: Correcting for measurement error. *Sociological Methods and Research, 14,* 447–472.

Jaccard, J., & Wan, C. K. (1995). Measurement error in the analysis of interaction effects between continuous predictors using multiple regression: Multiple indicator and structural equation approaches. *Psychological Bulletin, 117,* 348–357.

Jaccard, J., & Wan, C. K. (1996). *Lisrel approaches to interaction effects on multiple regression.* Beverly Hills, CA: Sage.

Jöreskog, K. G., & Yang, F. (1996). Nonlinear structural equation models: The Kenny–Judd model with interaction effects. In G. Marcoulides & R. Schumacker (Eds.), *Advanced structural equation modeling: Issues and techniques* (pp. 57–88). Mahwah, NJ: Lawrence Erlbaum Associates.

Kendall, M. G., & Stuart, A. (1969). *The advanced theory of statistics* (Vol. 1). London: Griffin.

Kenny, D. A., & Judd, C. M. (1984). Estimating the nonlinear and interactive effects of latent variables. *Psychological Bulletin, 96,* 201–210.

Ping, R. A., Jr. (1995). A parsimonious estimating technique for interaction and quadratic latent variables. *Journal of Marketing Research, 32,* 336–347.

Ping, R. A., Jr. (1996). Latent variable interaction and quadratic effect estimation: A two-step technique using structural equation analysis. *Psychological Bulletin, 119,* 166–175.

Tate, R. L. (1984). Limitations of centering for interactive models. *Sociological Methods and Research, 13,* 251–271.

APPENDIX A
Proof of Equation 14

By definition, the covariance between a variable and its product with another variable may be expressed as

$$C(x,xy) = E[[x - E(x)][xy - E(xy)]]. \tag{A1}$$

Using Equation 6 and expressing the variables as deviations from their means yields

$$C(x,xy) = E\Big[\big[\Delta x\big]\big[[E(x) + \Delta x][E(y) + \Delta(y)] - [E(x)E(y) + C(x,y)]\big]\Big]. \quad \text{(A2)}$$

$E(x)E(y)$ cancel. Distributing Δx and taking expectations gives

$$C(x,xy) = E(x)E(\Delta x\Delta y) + E(y)E(\Delta x)^2 + E[(\Delta x)^2(\Delta y)] - C(x,y)E(\Delta x). \quad \text{(A3)}$$

Because $E(\Delta x) = 0$, the expression becomes

$$C(x,xy) = E(x)E(\Delta x\Delta y) + E(y)E(\Delta x)^2 + E[(\Delta x)^2(\Delta y)], \quad \text{(A4)}$$

and finally

$$C(x,xy) = E(x)C(x,y) + E(y)\,V(x) + E[(\Delta x)^2(\Delta y)]. \quad \text{(A5)}$$

The odd moments of a normal distribution as well as the odd moments of a multivariate normal distribution are zero (Heise, 1986; Kendall & Stuart, 1969). Thus $E(\Delta x) = E[(\Delta x)^2(\Delta y)] = 0$.

APPENDIX B
SAS Program for the Monte Carlo Study

Following is the SAS code used to produce the estimates from artificial data reported in Table 9.1. PROC CALIS permits various ways to describe a structural equation model. We used the so-called “lineqs” method, which is derived from Bentler’s EQS model and tried, as far as possible, to give parameters names similar to those of the Jöreskog’s notation we used in the chapter. The Monte Carlo study was conducted using artificial samples derived from a theoretical covariance matrix and a theoretical mean vector. We included these in this appendix in place of a matrix and a vector computed from one of our artificial samples.

```
title "STRUCTURAL MODEL WITH INTERACTION AMONG THE LATENT VARIABLES";
title2 "THEORETICAL MEAN VECTOR AND COVARIANCE MATRIX";
options linesize=80;

data artif (type=cov);
 input _type_ $ _name_ $ y1-y5 x1-x9;
 cards;
n    .    2000  2000  2000   2000  2000  2000   2000   2000   2000  2000  2000
          2000  2000  2000
mean .      71.101  63.991   56.881   78.212   85.322    5.000    5.500    4.500
            6.000    7.200    6.600    7.000    6.300    5.600
cov  y1  112.580
```

```
          . . . . . . . . . . . . . .
cov y2   80.301  85.626
          . . . . . . . . . . . . .
cov y3   71.378  64.241   77.034
          . . . . . . . . . . . .
cov y4   98.145  88.331   78.516   127.910
          . . . . . . . . . . .
cov y5  107.068  96.361   85.654   117.774  162.116
          . . . . . . . . . .
cov x1    8.349   7.514    6.679     9.184   10.019   3.756
          . . . . . . . . .
cov x2    9.184   8.266    7.347    10.102   11.021   3.718    4.804
          . . . . . . . .
cov x3    7.514   6.763    6.011     8.266    9.017   3.042    3.346    4.090
          . . . . . . .
cov x4   10.060   9.054    8.048    11.065   12.071   1.014    1.116     .913
          5.408
          . . . . . .
cov x5   12.071  10.864    9.657    13.279   14.486   1.217    1.339    1.095
          5.841   8.510
          . . . . .
cov x6   11.065   9.959    8.852    12.172   13.279   1.116    1.227    1.004
          5.354   6.425    8.256
          . . .
cov x7   10.600   9.540    8.480    11.660   12.720   1.420    1.562    1.278
          1.136   1.363    1.249     7.361
          . .
cov x8    9.540   8.586    7.632    10.494   11.448   1.278    1.406    1.150
          1.022   1.227    1.124     5.963    6.765
          .
cov x9    8.480   7.632    6.784     9.328   10.176   1.136    1.249    1.022
           .909   1.090     .999     5.300    4.770   6.890
;

proc calis  data=artif ucov aug method=lsgls omethod=levmar dfred=1
            platcov residual toteff;
lineqs
 y1=       fETA1 + e1,
 y2= ly21 fETA1 + e2,
 y3= ly31 fETA1 + e3,
 y4= ly41 fETA1 + e4,
 y5= ly51 fETA1 + e5,
 x1=       fKSI1 + e6,
 x2= lx21 fKSI1 + e7,
 x3= lx31 fKSI1 + e8,
 x4=       fKSI2 + e9,
 x5= lx52 fKSI2 + e10,
 x6= lx62 fKSI2 + e11,
 x7=       fKSI3 + e12,
 x8= lx83 fKSI3 + e13,
 x9= lx93 fKSI3 + e14,
 fKSI1= k1 INTERCEP + d1,
```

```
 fKSI2= k2 INTERCEP + d2,
 fKSI3= k3 INTERCEP + d3,
 fKSI4= k4 INTERCEP + d4,
 fETA1= 50 INTERCEP + g11 fKSI1 + g12 fKSI2 + g13 fKSI3 + g14 fKSI4 + d5;
std
 e1-e5=  te1-te5,        /* residual variances of the endogenous
                            observed variables */
 e6-e14= td1-td9,        /* residual variances of the exogenous
                            observed variables */
 d1-d3=  ph11 ph22 ph33, /* variances of the latent exogenous
                            variables */
 d4=     ph44,           /* variance of KSI1*KSI2 */
 d5=     ps11;           /* residual variance of the latent
                            exogenous variable */
cov
 d1-d4= ph21 ph31 ph32 ph41 ph42 ph43;
                         /* covariances of the latent
                            exogenous variables */
/* nonlinear constraints */
ph44= (k1**2)*(ph22) + (k2**2)*(ph11) +2*k1*k2*ph21 + ph11*ph22 + (ph21**2);
                         /* variance of KSI1*KSI2 */
k4= k1*k2 + ph21;        /* mean of KSI1*KSI2 */
ph41= k1*ph21 + k2*ph11; /* covariance of KSI1 and KSI4 */
ph42= k2*ph21 + k1*ph22; /* covariance of KSI2 and KSI4 */
ph43= k1*ph32 + k2*ph31; /* covariance of KSI3 and KSI4 */
run;
```

10

Estimating Nonlinear Effects Using a Latent Moderated Structural Equations Approach

Karin Schermelleh-Engel
Andreas Klein
Helfried Moosbrugger
Johann Wolfgang Goethe University, Frankfurt am Main

In structural equation modeling (SEM), the latent variables in the equation are usually linearly related. But in some cases, theory suggests that in addition to the linear effects a product of the latent predictor variables may have an additional effect on a latent criterion variable. Kenny and Judd (1984) were one of the first who studied the methodological problems arising from nonlinearity. They formulated an elementary latent interaction model ("Kenny–Judd model") with two latent predictor variables, ξ_1 and ξ_2, each measured by two observed indicators (x_1, x_2, and x_3, x_4, respectively), and a latent criterion variable η measured by one observed indicator y (see Fig. 10.1). The elementary interaction model is characterized by the following structural equation (Equation 1) and measurement models (Equations 2 and 3):

$$\eta = \alpha + \gamma_1\xi_1 + \gamma_2\xi_2 + \gamma_3\xi_1\xi_2 + \zeta, \tag{1}$$

$$\begin{pmatrix} x_1 \\ x_2 \\ x_3 \\ x_4 \end{pmatrix} = \begin{pmatrix} 1 & 0 \\ \lambda_{21} & 0 \\ 0 & 1 \\ 0 & \lambda_{42} \end{pmatrix} \begin{pmatrix} \xi_1 \\ \xi_2 \end{pmatrix} + \begin{pmatrix} \delta_1 \\ \delta_2 \\ \delta_3 \\ \delta_4 \end{pmatrix}, \tag{2}$$

$$y = \eta. \tag{3}$$

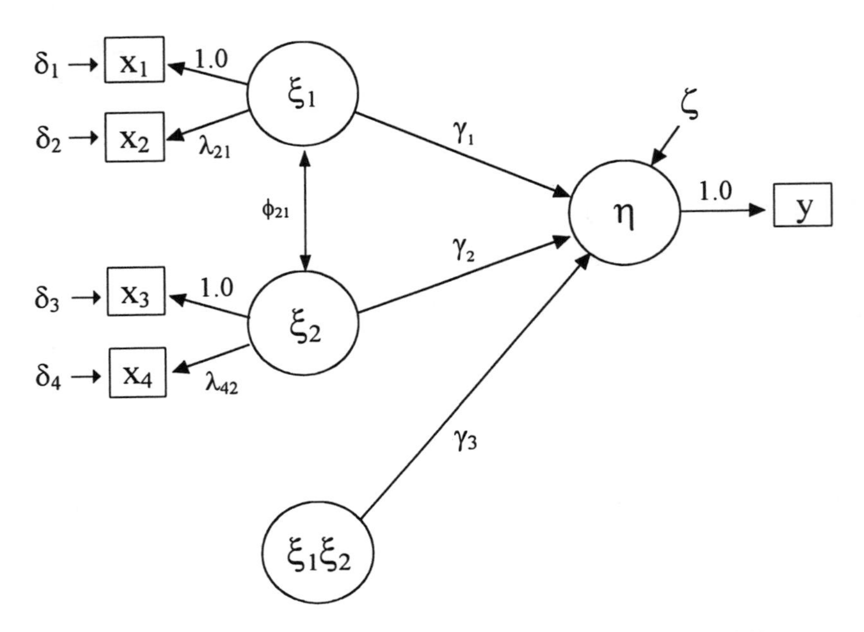

FIG. 10.1. Elementary interaction model with two latent predictor variables ξ_1, ξ_2, one latent criterion variable η, and a latent product variable $\xi_1\xi_2$.

The following assumptions are made:

- $x_1, \ldots, x_4$ are multivariate normal with zero means.
- ξ_1 and ξ_2 are bivariate normal with zero means.
- $\delta_1, \ldots, \delta_4$ are normal with zero means.
- ζ is normal with zero mean.
- δ_i is independent of δ_j for $i \neq j$ ($i = 1, \ldots, 4; j = 1, \ldots, 4$).
- δ_i is independent of ξ_k for $i = 1, \ldots, 4$ and $k = 1, 2$.
- ζ is independent of δ_i and ξ_k for $i = 1, \ldots, 4$ and $k = 1, 2$.

One problem with analyzing latent interaction models in LISREL is that a measurement model for the latent product term $\xi_1\xi_2$ has to be defined. Kenny and Judd (1984) suggested using four product variables (x_1x_3, x_1x_4, x_2x_3, and x_2x_4) as a measurement model for the latent product term $\xi_1\xi_2$. Although they did not include the constant intercept term α in Equation 1, this should normally be done as α may still be nonzero even if y, ξ_1, ξ_2, and ζ have zero means (Jöreskog & Yang, 1997).

Latent interaction models involve nonlinear structural relationships between latent variables, and the distributions resulting from nonlinear terms in the structural equation are known to be non-normal (Jöreskog & Yang, 1996). Relative to the size of the structural equation coefficients, for example, skewness and kurtosis of the dependent variables indicate substantial deviation from normality.

For an estimation method, this fact has two possible consequences: First, if an estimation procedure is used under the assumption of normally distributed indicator variables, the robustness for interaction models should be investigated thoroughly. For inferential statistics, the possible bias of the estimated standard errors can become critical in the method's performance. Second, estimation procedures that are asymptotically distribution-free (e.g., LISREL-WLSA) must be tested with regard to their power and efficiency. Because they do not make distributional assumptions, they ignore the specific stochastic structure implied by product terms. Simulation studies can decide whether this lowers the method's efficiency, which might become very low when sample size decreases.

Because of the problems with non-normality in latent interaction models, the new estimation method *latent moderated structural equations* (LMS)[1] was developed (Klein & Moosbrugger, 1997; Klein, Moosbrugger, Schermelleh-Engel, & Frank, 1997; Moosbrugger, Schermelleh-Engel, & Klein, 1997). Based on a stochastic analysis of the multivariate density function of indicator variables, LMS implements an iterative maximum likelihood (ML) estimation for the model parameters. The new method takes the non-normality of the variables explicitly into account and does not require the forming of any product variables.

In this chapter, we compare the performance of LMS with methods used for the analysis of latent interaction effects in recent research in a simulation study: LISREL-ML (maximum likelihood), LISREL-WLSA (weighted least squares based on the augmented moment matrix; Jöreskog & Yang, 1996, 1997; Yang Jonsson, 1997), and 2SLS (two-stage least squares estimator; Bollen, 1995, 1996; Bollen & Paxton, chap. 6). Jöreskog and Yang (1996, 1997) proposed LISREL-WLSA as the asymptotically optimal method in LISREL, because it provides asymptotically correct standard errors for the estimates. Because WLSA requires large sample sizes for establishing the asymptotic properties, they suggested that LISREL-ML could be used in many cases, although, from a theoretical standpoint, the assumption of multivariate normality is violated in LISREL-ML.

As least squares estimation methods are not based on this assumption, they have been proposed as an alternative. Bollen (1995, 1996) developed a general method that makes use of the 2SLS for structural equation models with nonlinear functions of latent variables. This general approach permits any number of equations, allows other nonlinear functions of the latent or

[1]The LMS method was developed as part of the doctoral thesis of Andreas Klein.

observed variables, and applies to the measurement model as well as to the latent variable model.

The content of this chapter is as follows: After a description of the main concept and procedures of LMS, we explain the purpose and results of the simulation study. The implementation of the elementary interaction model for LISREL-ML, LISREL-WLSA, and 2SLS is summarized. Though the asymptotic estimation properties of LMS and the other methods are known, the simulation study reveals the methods' performance for finite samples under specified conditions.

LATENT MODERATED STRUCTURAL EQUATIONS

In view of non-normality problems with latent interaction models, a stochastic analysis of the density functions involved by interaction models could provide a theoretical background for the development of efficient estimation methods and a better understanding of the major problems associated with latent interaction. The LMS approach is based on a derivation of the multivariate density function of the indicator variables, and it takes the type of non-normal distribution implied by interaction models explicitly into account (a prototype version of LMS is described in Klein et al., 1997). It represents the density function as a finite mixture of multivariate normal densities. With the estimation maximation (EM) algorithm (Dempster, Laird, & Rubin, 1977) adapted to the mixture density, LMS provides an iterative ML estimation of all model parameters. Because the non-normal density function of the indicator vector is explicitly given in LMS, the standard errors of the LMS estimators can be numerically calculated with the Fisher information matrix. At least, LMS estimators have the properties given by general ML estimation theory: They are consistent, asymptotically unbiased, asymptotically most efficient, and asymptotically normally distributed.

The finite sample properties of the estimators can be investigated via simulation studies (see the section Properties of LMS Versus LISREL and 2SLS). Because the density functions analyzed by LMS are a nonlinear extension of densities involved by multilinear models, LMS cannot yet be implemented with standard statistical software for latent variable models. So far, the LMS algorithm is written in Delphi Pascal Code for several models with one or two latent interaction effects.

The main concept and procedures of the LMS approach are explained with reference to the elementary interaction model (Equations 1 to 3). Because the derivation of the mixture density and the adaptation to the EM algorithm is rather technical, the details are omitted in this chapter. A version of LMS applicable to more complex interaction models including implementation details and a general expression in matrix notation form is given in a separate article (Klein & Moosbrugger, 1997).

Stochastic Analysis of Density Function

The elementary interaction model (Equations 1 to 3) has a five-dimensional indicator vector $(\mathbf{x}, \mathbf{y}) = (x_1, \ldots, x_4, y)$. The indicator $\mathbf{x}$ (4×1) of latent independent variables ξ (2×1) is normally distributed, whereas the indicator $\mathbf{y}$ of latent dependent variable η is not because of the product term $\xi_1\xi_2$ in the structural equation. An analysis of the distribution of vector $(\mathbf{x}, \mathbf{y})$ shows that $(\mathbf{x}, \mathbf{y})$ follows a continuous mixture of multivariate normal distributions with a standardized normal variate $\mathbf{z}$ as a mixing variable. The mean vectors and covariance matrices of the mixture components depend on realizations of the mixing variable $\mathbf{z}$ and the model parameters. Then, the density function $f(\mathbf{x} = x, \mathbf{y} = y)$ of indicator vector $(\mathbf{x}, \mathbf{y})$ can be expressed in the form

$$f(\mathbf{x} = x,\ \mathbf{y} = y) = \int_{-\infty}^{\infty} \varphi_{0,1}(z) \cdot \varphi_{\mu(z),\Sigma(z)}(x, y)\, dz, \tag{4}$$

where $\varphi_{0,1}(z)$ is the standardized normal density of the mixing variable $\mathbf{z}$:

$$\varphi_{0,1}(z) = \frac{1}{\sqrt{2\pi}} \exp(-\frac{1}{2}z^2). \tag{5}$$

The expression $\varphi_{\mu(z),\, \Sigma(z)}(x, y)$ is a five-dimensional multivariate normal density, and the coefficients of its mean vector $\mu(z)$ (5×1) and covariance matrix $\Sigma(z)$ (5×5) are functions of the model parameters and z, the realizations of the mixing variable $\mathbf{z}$. To explicate these functions, we create a partitioned mean vector and covariance matrix:

$$\mu(z) = \begin{bmatrix} \mu_x(z) \\ \mu_y(z) \end{bmatrix},\ \Sigma(z) = \begin{bmatrix} \Sigma_{xx}(z) & \Sigma_{xy}(z) \\ \Sigma_{xy}(z)' & \Sigma_{yy}(z) \end{bmatrix}. \tag{6}$$

Further, let $\mathbf{\Phi} = \mathbf{AA}'$ be the Cholesky decomposition of the covariance matrix $\mathbf{\Phi}$ (2×2) of ξ with lower triangular matrix $\mathbf{A}$ (2×2):

$$\begin{pmatrix} \phi_{11} & \phi_{12} \\ \phi_{21} & \phi_{22} \end{pmatrix} = \begin{pmatrix} a_{11} & 0 \\ a_{21} & a_{22} \end{pmatrix} \begin{pmatrix} a_{11} & 0 \\ a_{21} & a_{22} \end{pmatrix}'. \tag{7}$$

Then, as a result of the density analysis, the subvectors $\mu_x(z)$, $\mu_y(z)$ and submatrices $\Sigma_{xx}(z)$, $\Sigma_{xy}(z)$, $\Sigma_{yy}(z)$ of the partitions (Equation 6) have the form

$$\mu_x(z) = \begin{pmatrix} a_{11}z \\ \lambda_{21}a_{11}z \\ a_{21}z \\ \lambda_{42}a_{21}z \end{pmatrix}, \tag{8}$$

$$\mu_y(z) = \left(\alpha + (\gamma_1 a_{11} + \gamma_2 a_{21})z + \gamma_3 a_{11}a_{21}z^2\right), \tag{9}$$

$$\Sigma_{xx}(z) = \begin{pmatrix} \theta_{11} & 0 & 0 & 0 \\ 0 & \theta_{22} & 0 & 0 \\ 0 & 0 & a_{22}^2 + \theta_{33} & \lambda_{42}a_{22}^2 \\ 0 & 0 & \lambda_{42}a_{22}^2 & \lambda_{42}^2 a_{22}^2 + \theta_{44} \end{pmatrix}, \tag{10}$$

$$\Sigma_{xy}(z) = \begin{pmatrix} 0 \\ 0 \\ \gamma_2 a_{22}^2 + \gamma_3 a_{11}a_{22}^2 z \\ \lambda_{42}(\gamma_2 a_{22}^2 + \gamma_3 a_{11}a_{22}^2 z) \end{pmatrix}, \tag{11}$$

$$\Sigma_{yy}(z) = (\gamma_2^2 a_{22}^2 + 2\gamma_2\gamma_3 a_{11}a_{22}^2 z\ \gamma_3^2 a_{11}^2 a_{22}^2 z^2 + \psi_{11})\ . \tag{12}$$

Note that $\Sigma_{xx}(z)$ is independent from z, whereas $\mu_x(z)$ and $\Sigma_{xy}(z)$ depend only linearly on z. If the interaction model holds and γ_3 is different from zero, subvector $\mu_y(z)$ and submatrix $\Sigma_{yy}(z)$ depend on z quadratically. If this is the case, the integral of the mixture density (Equation 4) cannot be solved analytically. Instead, it is approximated by Hermite quadrature formulas for numerical integration of weighted integrands (Isaacson & Keller, 1966). The approximation of Equation 4 yields a weighted sum

$$f(\mathbf{x} = x, \mathbf{y} = y) = \sum_{j=1}^{M} \rho_j \varphi_{\mu(2^{1/2}v_j),\Sigma(2^{1/2}v_j)}(x,y) \tag{13}$$

so that, up to a remainder negligible in size, f can be expressed as a finite mixture of normal densities. The weights ρ_j for the approximation are calculated by

$$\rho_j = \frac{w_j}{\sqrt{\pi}}, \tag{14}$$

where w_j and v_j (see Equation 13) are the weights and abscissas listed for Hermite quadrature formulas of order M (see Abramowitz & Stegun, 1971). Note that the weights and abscissas calculated for the finite mixture are derived analytically, and they are independent from model parameters. Additionally, an important property of the finite mixture is that the parameters of the mixture components in Equation 13 are not independent, which involves a more complicated situation for ML estimation than finite mixtures with independent components. The higher M is chosen for the quadrature, the more exact the approximation is. For the elementary interaction model examined here, for example, M = 16 provides a sufficiently exact approximation of the density.

ML Estimation

A common approach to the ML estimation of parametric models with mixture densities is the application of the EM algorithm (Redner & Walker, 1984). Under fairly general conditions, the EM algorithm provides an iterative estimation procedure that converges to a ML estimation of the parameters (Dempster et al., 1977). For the estimation of the elementary interaction model (Equations 1 to 3), the maximum number of EM iterations was limited to 50. Every iteration step consists of two subordinate steps, the estimation step and the maximation step.

Let θ be the parameter vector of the elementary interaction model (Equations 1 to 3), and let $\theta^{(0)}$ be a vector of starting values for the parameters. We describe the kth iteration step of the EM algorithm for LMS. At the beginning of step k the current value for θ is $\theta^{(k-1)}$. Then the estimation step is the calculation of a $(N \times \mathrm{M})$ matrix $\mathbf{P}^{(k)} = [p^{(k)}(\mathrm{j}|\mathbf{x}_\mathrm{i}, \mathbf{y}_\mathrm{i})]_{\mathrm{i}=1\ldots\mathrm{N},\mathrm{j}=1\ldots\mathrm{M}}$ of density ratios:

$$p^{(k)}(\mathrm{j}|\mathbf{x}_\mathrm{i},\mathbf{y}_\mathrm{i}) = \frac{\rho_\mathrm{j}\varphi_{\mu(2^{1/2}v_\mathrm{j}),\Sigma(2^{1/2}v_\mathrm{j})}(\mathbf{x}_\mathrm{i},\mathbf{y}_\mathrm{i})}{f(\mathbf{x}_\mathrm{i},\mathbf{y}_\mathrm{i})} \tag{15}$$

where $(\mathbf{x}_\mathrm{i}, \mathbf{y}_\mathrm{i})$ is the ith row of the indicators' data matrix, and the current parameter values of $\theta^{(k-1)}$ are used for calculation of mean vectors $\mu(2^{1/2}v_\mathrm{j})$ and covariance matrices $\Sigma(2^{1/2}v_\mathrm{j})$ occurring in the nominator or denominator of $p^{(k)}(\mathrm{j}|\mathbf{x}_\mathrm{i}, \mathbf{y}_\mathrm{i})$. Here, in contrast to covariance structure analysis, LMS needs the raw data for computation. The maximation step is the calculation of the new parameter vector $\theta^{(k)}$ as the value of θ, which maximizes a sum weighted by the density ratios computed in the estimation step:

$$\theta^{(k)} = \arg\max_{\theta} \left\{ \sum_{i=1}^{N} \sum_{j=1}^{M} p^{(k)}(j|\boldsymbol{x}_i,\boldsymbol{y}_i) \ln \varphi_{\mu(2^{1/2}v_j),\Sigma(2^{1/2}v_j)}(\boldsymbol{x}_i,\boldsymbol{y}_i) \right\} \tag{16}$$

where $p^{(k)}(\mathrm{j}|\mathbf{x}_\mathrm{i}, \mathbf{y}_\mathrm{i})$ are the calculated density ratios, and $\mu(2^{1/2}v_\mathrm{j})$, $\Sigma(2^{1/2}v_\mathrm{j})$ on the right-hand side are functions of θ. For the calculation of $\theta^{(k)}$, numerical computation of partial derivatives and the single-step iteration method (Isaacson & Keller, 1966; Schwarz, 1993) for solving systems of nonlinear

(Isaacson & Keller, 1966; Schwarz, 1993) for solving systems of nonlinear equations are used in LMS. The sequence $[\theta^{(k)}]_{k=0,1,2,\ldots}$ converges to a ML estimation of $\hat{\theta}$ of θ.

Because the EM algorithm provides ML estimates of all parameters and the structure of the density of indicator vector ($\mathbf{x}$, $\mathbf{y}$) is known in LMS, the standard errors can be numerically evaluated with the Fisher information matrix, as follows from general ML estimation theory. Because the implementation of this part is very technical and only a straightforward application of numerical methods, the details are not given in this chapter. Though not yet implemented in the LMS algorithm but straightforward in application, a model test for interaction hypotheses could be carried out by calculating the coefficients for the likelihood ratio test statistic.

After all, by analyzing and utilizing the density function of indicator variables, LMS is still a computationally intensive, iterative estimation procedure tailored for latent interaction models. Unlike LISREL or 2SLS (Bollen, 1996) procedures, it needs the raw data for estimation and does not require the forming of any products of indicator variables. For the analysis of empirical data, LMS assumes the indicators of the exogenous latent variables ξ to be normally distributed, which should be verified before using LMS. As a common recommendation valid for all methods, the application of LMS should not be executed routinely. In general, there should be theoretical reasons for setting up a latent interaction model.

PROPERTIES OF LMS VERSUS LISREL AND 2SLS

Planning a Simulation Study

The performance of the LMS estimation technique for finite samples can be checked in Monte Carlo simulation studies with artificial data sets. Important properties of an estimation method are bias, efficiency of its parameter estimators, and, for inferential statistics, the bias of the estimated standard errors. Estimation methods designed for interaction models should further be tested as to whether they are robust against higher degrees of non-normality in the distributions caused by substantial interaction effects in the structural equation. For the plan of a simulation study, the models and sample sizes should be chosen in such a way that the characteristics of an estimation method concerning these properties can be detected. Besides LMS, the estimation methods LISREL-ML, LISREL-WLSA, and 2SLS (Bollen, 1995, 1996) were selected as candidates for the study.

We chose the elementary interaction model (Equations 1 to 3) and simulated artificial data sets with varying parameter values for γ_3 and varying sample size N. For data generation, the 14 model parameters were fixed to

TABLE 10.1
Parameter List of the Elementary Interaction Model

Parameter	*Value*	*Parameter*	*Value*
γ_1	0.20	ϕ_{22}	0.64
γ_2	0.40	λ_{21}	0.60
γ_3	0.30 (A) / 0.70 (B) / 1.50 (C)	λ_{42}	0.70
α	1.00	θ_{11}	0.51
ψ_{11}	0.20	θ_{22}	0.64
ϕ_{11}	0.49	θ_{33}	0.36
φ_{21}	0.2352	θ_{44}	0.51

the following values (see Table 10.1), which are taken from Jöreskog and Yang (1996, 1997).

Parameter γ_3, the size of the latent interaction effect, was varied by $\gamma_3 = 0.30$ (Model A), $\gamma_3 = 0.70$ (Model B), and $\gamma_3 = 1.50$ (Model C). Sample size N was varied by $N = 200$, $N = 400$, and $N = 800$. There are nine possible combinations of γ_3 and N. Using the PRELIS program (Jöreskog & Sörbom, 1996b), 500 data sets for each of the nine combinations were generated (see Table 10.2), hereafter referred to as data packages. Every data set is represented by a $(N \times 5)$ data matrix of the indicator vector $(x_1, \ldots, x_4, y)$.

Each data package was analyzed by each method separately, providing nine Monte Carlo studies for each method. For each of the nine data packages, each method computed 500 estimations (except nonconvergent solutions) for every parameter. The distributions of the parameter estimates were examined by calculation of means and standard deviations (MC-SDs). Also, the means of the estimated standard errors (Est-SDs) were computed for every data package and method.

For detection of bias, the number of data sets for each data package was chosen to be 500, so that for an unbiased estimator, the mean of the estimates should not deviate substantially from its true value. The relatively low value

TABLE 10.2
Design of Data Generation for Nine Monte Carlo Studies to Analyze the Elementary Interaction Model With True Parameters Taken From Jöreskog and Yang (1996)

	Interaction Effect		
Sample Size	$\gamma_3 = 0.30$ *(Model A)*	$\gamma_3 = 0.70$ *(Model B)*	$\gamma_3 = 1.50$ *(Model C)*
$N = 200$	500 data sets	500 data sets	500 data sets
$N = 400$	500 data sets	500 data sets	500 data sets
$N = 800$	500 data sets	500 data sets	500 data sets

$\gamma_3 = 0.30$ (Model A) was chosen to test the methods' power for detecting low interaction effects, whereas the value $\gamma_3 = 1.50$ (Model C) was selected to reveal robustness problems. The sample size $N = 200$ was selected to examine the limits of the methods' power or robustness for small samples typical of experiments in the behavioral or social sciences.

Implementation of Alternative Methods

Implementation of LISREL-ML. For the LISREL implementation of the Kenny–Judd model, the extended LISREL model (LISREL with mean structures; Jöreskog & Sörbom, 1989, 1996a) with four product variables is used. The equations are written in general LISREL notation. The LISREL-ML approach estimates the parameter vector θ by minimizing the fit function

$$F(\theta) = \log\|\Sigma\| + \mathrm{tr}(S\Sigma^{-1}) - \log\|S\| - k + (\bar{z} - \mu)'\Sigma^{-1}(\bar{z} - \mu), \tag{17}$$

where

- $\bar{z}$ is the empirical mean vector, S the empirical covariance matrix,
- μ is the population mean vector, Σ the population covariance matrix,
- k is the number of indicator variables in z.

The fit function is derived from the ML principle based on the assumption that the k indicator variables in z have a multinormal distribution. The assumption of multivariate normality is violated by using the non-normal indicator y and product variables in the elementary interaction model. As Yang Jonsson (1997) stated, the bias of the parameter estimates is not serious for $N \geq 400$, but standard errors are severely underestimated.

The model is easily estimated with LISREL 8 using nonlinear constraints (Jöreskog & Sörbom, 1996a). The model implementation follows the definitions given by Yang Jonsson (1997) with the simplification that all regression constants ($\tau_1, \ldots, \tau_4$) of the measurement model for $x_1, \ldots, x_4$ are fixed to zero. Because the indicators $x_1, \ldots, x_4$ are centered in the simulation study, this simplification is possible. The structural equation of the elementary interaction model is

$$\eta = \alpha + (\gamma_1 \;\; \gamma_2 \;\; \gamma_3)\begin{pmatrix} \xi_1 \\ \xi_2 \\ \xi_1\xi_2 \end{pmatrix} + \zeta \tag{18}$$

with measurement model for the latent dependent variable η:

$$y = \eta. \tag{19}$$

The measurement model for the latent predictor variables is given by

$$\begin{pmatrix} x_1 \\ x_2 \\ x_3 \\ x_4 \\ x_1x_3 \\ x_1x_4 \\ x_2x_3 \\ x_2x_4 \end{pmatrix} = \begin{pmatrix} 1 & 0 & 0 \\ \lambda_{21} & 0 & 0 \\ 0 & 1 & 0 \\ 0 & \lambda_{42} & 0 \\ 0 & 0 & 1 \\ 0 & 0 & \lambda_{42} \\ 0 & 0 & \lambda_{21} \\ 0 & 0 & \lambda_{21}\lambda_{42} \end{pmatrix} \begin{pmatrix} \xi_1 \\ \xi_2 \\ \xi_1\xi_2 \end{pmatrix} + \begin{pmatrix} \delta_1 \\ \delta_2 \\ \delta_3 \\ \delta_4 \\ \delta_5 \\ \delta_6 \\ \delta_7 \\ \delta_8 \end{pmatrix}. \tag{20}$$

The mean vector κ and covariance matrix Φ of latent predictor variables and error covariance matrix are Θ_δ

$$\kappa = \begin{pmatrix} 0 \\ 0 \\ \phi_{21} \end{pmatrix}, \qquad \Phi = \begin{pmatrix} \phi_{11} & & \\ \phi_{21} & \phi_{22} & \\ 0 & 0 & \phi_{11}\phi_{22} + \phi_{21}^2 \end{pmatrix}, \tag{21}$$

$$\Theta_\delta = \begin{pmatrix} \theta_{11} & & & & & & & \\ 0 & \theta_{22} & & & & & & \\ 0 & 0 & \theta_{33} & & & & & \\ 0 & 0 & 0 & \theta_{44} & & & & \\ 0 & 0 & 0 & 0 & \theta_{55} & & & \\ 0 & 0 & 0 & 0 & \theta_{65} & \theta_{66} & & \\ 0 & 0 & 0 & 0 & \theta_{75} & 0 & \theta_{77} & \\ 0 & 0 & 0 & 0 & 0 & \theta_{86} & \theta_{87} & \theta_{88} \end{pmatrix}, \tag{22}$$

where $\theta_{65} = \lambda_{42}\phi_{22}\theta_{11}$, $\theta_{75} = \lambda_{21}\phi_{11}\theta_{33}$, $\theta_{86} = \lambda_{21}\phi_{11}\theta_{44}$, $\theta_{87} = \lambda_{42}\phi_{22}\theta_{22}$, $\theta_{55} = \phi_{11}\theta_{33} + \phi_{22}\theta_{11} + \theta_{11}\theta_{33}$, $\theta_{66} = \phi_{11}\theta_{44} + \lambda_{42}^2\phi_{22}\theta_{11} + \theta_{11}\theta_{44}$, $\theta_{77} = \lambda_{21}^2\phi_{11}\theta_{33} + \phi_{22}\theta_{22} + \theta_{22}\theta_{33}$, and $\theta_{88} = \lambda_{21}^2\phi_{11}\theta_{44} + \lambda_{42}^2\phi_{22}\theta_{22} + \theta_{22}\theta_{44}$. The LISREL-ML input file is given in Appendix A.

Implementation of LISREL-WLSA. LISREL-WLSA is applied with four products of indicators used to measure the latent product term. The WLSA approach estimates θ by minimizing the fit function

$$F(\theta) = (a - \alpha)' \, W_\alpha^- \, (a - \alpha), \tag{23}$$

where

- $A = (a_{ij}) = \begin{pmatrix} S + \bar{z}\bar{z}' & \\ \bar{z}' & 1 \end{pmatrix}$ (elements in the lower half of the matrix),
- $Y = (\alpha_{ij}) = \begin{pmatrix} \Sigma + \mu\mu' & \\ \mu' & 1 \end{pmatrix}$ (elements in the lower half of the matrix),
- $\bar{z}$ is the empirical mean vector, S the empirical covariance matrix,
- μ is the population mean vector, Σ the population covariance matrix,
- W_α is a consistent estimate of the covariance matrix Ω_α of a,
- W_α^- is the Moore–Penrose generalized inverse of W_α.

The sample augmented moment matrix A is the matrix of sample moments about zero for the vector z augmented by a variable of constant values 1 for all cases. For a more detailed description of WLSA see Yang Jonsson (1997).

As proposed by Jöreskog and Yang (1996), we choose a so-called η model as the adequate LISREL model for WLSA. Again, the regression constants of the measurement models are not included, because the indicator variables $x_1, \ldots, x_4$ are centered. The structural equation model is (Yang Jonsson, 1997)

$$\begin{pmatrix} \eta_1 \\ \eta_2 \\ \eta_1\eta_2 \\ 1 \end{pmatrix} = \begin{pmatrix} 0 \\ 0 \\ \psi_{21} \\ 1 \end{pmatrix} 1 + \begin{pmatrix} \eta_1 \\ \eta_2 \\ \eta_1\eta_2 - \psi_{21} \\ 0 \end{pmatrix}. \tag{24}$$

The measurement model is defined as follows:

$$\begin{pmatrix} y_1 \\ y_2 \\ y_3 \\ y_4 \\ y_5 \\ y_2y_4 \\ y_2y_5 \\ y_3y_4 \\ y_3y_5 \end{pmatrix} = \begin{pmatrix} \lambda_{11} & \lambda_{12} & \lambda_{13} & \lambda_{14} \\ 1 & 0 & 0 & 0 \\ \lambda_{31} & 0 & 0 & 0 \\ 0 & 1 & 0 & 0 \\ 0 & \lambda_{52} & 0 & 0 \\ 0 & 0 & 1 & 0 \\ 0 & 0 & \lambda_{52} & 0 \\ 0 & 0 & \lambda_{31} & 0 \\ 0 & 0 & \lambda_{31}\lambda_{52} & 0 \end{pmatrix} \begin{pmatrix} \eta_1 \\ \eta_2 \\ \eta_1\eta_2 \\ 1 \end{pmatrix} + \begin{pmatrix} \varepsilon_1 \\ \varepsilon_2 \\ \varepsilon_3 \\ \varepsilon_4 \\ \varepsilon_5 \\ \varepsilon_6 \\ \varepsilon_7 \\ \varepsilon_8 \\ \varepsilon_9 \end{pmatrix}, \tag{25}$$

where $y_2, \ldots, y_5$ equal $x_1, \ldots, x_4$, $\lambda_{11}, \ldots, \lambda_{13}$ equal $\gamma_1, \ldots, \gamma_3$, and λ_{14} equals the intercept term α in the previous LISREL-ML model. The parameter matrices ψ and Θ_ε include the following elements:

$$\psi = \begin{pmatrix} \psi_{11} & & & \\ \psi_{21} & \psi_{22} & & \\ 0 & 0 & \psi_{11}\psi_{22} + \psi_{21}^2 & \\ 0 & 0 & 0 & 0 \end{pmatrix}, \quad (26)$$

$$\Theta_\varepsilon = \begin{pmatrix} \theta_{11} & & & & & & & & \\ 0 & \theta_{22} & & & & & & & \\ 0 & 0 & \theta_{33} & & & & & & \\ 0 & 0 & 0 & \theta_{44} & & & & & \\ 0 & 0 & 0 & 0 & \theta_{55} & & & & \\ 0 & 0 & 0 & 0 & 0 & \theta_{66} & & & \\ 0 & 0 & 0 & 0 & 0 & \theta_{76} & \theta_{77} & & \\ 0 & 0 & 0 & 0 & 0 & \theta_{86} & 0 & \theta_{88} & \\ 0 & 0 & 0 & 0 & 0 & 0 & \theta_{97} & \theta_{98} & \theta_{99} \end{pmatrix}, \quad (27)$$

where $\theta_{76} = \lambda_{52}\psi_{22}\theta_{22}$, $\theta_{86} = \lambda_{31}\psi_{11}\theta_{44}$, $\theta_{97} = \lambda_{31}\psi_{11}\theta_{55}$, $\theta_{98} = \lambda_{52}\psi_{22}\theta_{33}$, $\theta_{66} = \psi_{11}\theta_{44} + \psi_{22}\theta_{22} + \theta_{22}\theta_{44}$, $\theta_{77} = \psi_{11}\theta_{55} + \lambda_{52}^2\psi_{22}\theta_{22} + \theta_{22}\theta_{55}$, $\theta_{88} = \lambda_{31}^2\psi_{11}\theta_{44} = \psi_{22}\theta_{33} + \theta_{33}\theta_{44}$, and $\theta_{99} = \lambda_{31}^2\psi_{11}\theta_{55} + \lambda_{52}^2\psi_{22}\theta_{33} + \theta_{33}\theta_{55}$. The LISREL-WLSA input file is given in Appendix B.

Implementation of 2SLS. Bollen (1995, 1996) and Bollen and Paxton (chap. 6) developed a 2SLS estimation method for latent variable models (structural equation models). The 2SLS approach proposes a noniterative estimation procedure that provides consistent parameter estimators and allows significance tests by calculating the asymptotic covariance matrix of the estimators. The method can be implemented easily with standard statistical software, which is of importance for latent interaction models, and the indicator variables are allowed to originate from non-normal distributions.

For the estimation of the regression coefficients of the latent variable model, the 2SLS algorithm requires two sets of observed variables: scaling variables (SV) and instrumental variables (IV). The SVs a priori have factor loadings of 1 in their measurement equations. The IVs must be correlated with the SVs, but uncorrelated with the measurement errors of the SVs. Furthermore, there must be at least as many IVs as the number of explanatory variables on the right-hand side of the structural equation (for details about the conditions on IVs, see Bollen, 1996). In the first stage of 2SLS, one selects the SVs that scale the explanatory variables of the structural equation. A regression of these SVs on the IVs is executed, providing coefficient estimates for the regression equation. Then, one calculates the regression values predicted for the selected SVs by the IVs in that regression. The second stage consists of an OLS regression of the SVs for the latent dependent random vector η on the regression values predicted for the selected SVs in Stage 1. Thus, using the predicted regression values from Stage 1 for a regression in Stage 2, the 2SLS method provides a consistent estimator of the regression coefficients in the structural equation.

We explain the application of 2SLS to the elementary interaction model (Equations 1 to 3), as proposed in Bollen (1995). Although the 2SLS method, in a generalized version, gives estimates of all model parameters, our application of 2SLS, for the purpose of testing LMS, is restricted to the estimation of the structural equation's regression coefficients α, γ_1, γ_2, and γ_3, as proposed by Bollen.

In a preliminary step of 2SLS, sets of appropriate SVs and IVs must be determined for the elementary interaction model. The indicator y is a SV for η, and x_1, x_3, and x_1x_3 are SVs for the explanatory variables ξ_1, ξ_2, and $\xi_1\xi_2$ on the right-hand side of Equation 1. Let $\mathbf{y}$ ($N \times 1$) contain the N values of y. Let $\mathbf{Z}$ ($N \times 4$) contain 1's in the first column and the N rows of the SVs x_1, x_3, and x_1x_3, respectively.

Let $\mathbf{A} = (\alpha, \gamma_1, \gamma_2, \gamma_3)'$ be the vector of regression coefficients in the structural equation. Then the model equations (Equations 1 to 3) imply an equation for the SVs

$$\mathbf{y} = \mathbf{Z}\,\mathbf{A} + \mathbf{u}, \tag{28}$$

where the error term $\mathbf{u}$ ($N \times 1$) contains the values of $\zeta - \gamma_1\delta_1 - \gamma_2\delta_2 - \gamma_3[\xi_1\delta_3 + \xi_2\delta_1 + \delta_1\delta_3]$. Now the variables x_2, x_4, and x_2x_4 form IVs: They are correlated with the SVs x_1, x_3, and x_1x_3, and the error terms of x_2, x_4, and x_2x_4 are uncorrelated with the errors of x_1, x_3, and x_1x_3. Additionally, a fourth IV $\hat{x}_1\hat{x}_3$ can be formed by regressing x_1 and x_3 on the three IVs x_2, x_4, and x_2x_4, and forming predicted variables $\hat{x}_1$ and $\hat{x}_3$ (Bollen, 1995).

In the first stage of 2SLS, we form the IVs' data matrix $\mathbf{V}$ ($N \times 5$), which contains 1's in the first column and the N rows of IVs x_2, x_4, x_2x_4, and $\hat{x}_1\hat{x}_3$, respectively. Then the OLS estimation for the regression of $\mathbf{Z}$ on $\mathbf{V}$ entails

$$(\mathbf{V}'\mathbf{V})^{-1}\,\mathbf{V}'\mathbf{Z}, \tag{29}$$

and the matrix $\hat{\mathbf{Z}}$ with predicted regression values for $\mathbf{Z}$ becomes

$$\hat{\mathbf{Z}} = \mathbf{V}(\mathbf{V}'\mathbf{V})^{-1}\mathbf{V}'\mathbf{Z}. \tag{30}$$

In the second stage, an OLS regression of **y** on $\hat{\mathbf{Z}}$ provides coefficient estimates for **A**:

$$\hat{\mathbf{A}} = (\hat{\mathbf{Z}}'\,\hat{\mathbf{Z}})^{-1}\hat{\mathbf{Z}}'\,\mathbf{y}. \tag{31}$$

The distribution of $\hat{\mathbf{A}}$ is normal with asymptotic covariance matrix (Bollen, 1996)

$$\text{acov}(\hat{\mathbf{A}}) = \hat{\sigma}_u^2(\hat{\mathbf{Z}}'\,\hat{\mathbf{Z}})^{-1}, \tag{32}$$

where $\hat{\sigma}_u^2 = {}^1\!/\!_N(\mathbf{y} - \mathbf{Z}\hat{\mathbf{A}})'\,(\mathbf{y} - \mathbf{Z}\hat{\mathbf{A}})$. Now the standard errors are calculated by taking square roots of diagonal elements of acov($\hat{\mathbf{A}}$). The properties of the 2SLS estimators are large-sample properties, and the simulation study (see the section Finite Sample Properties of LMS Estimators) illustrates the performance of the estimator $\hat{\mathbf{A}}$ and acov($\hat{\mathbf{A}}$) in finite samples.

Estimation Results

The results of the simulation study with means of parameter estimates (M) and standard deviations of parameter estimates (MC-SD) are completely listed in Appendix C. The MC-SD values represent the "true" standard error of the parameter estimators. For each data set analyzed in the simulation study by one of the four methods, an estimated standard error for the parameter estimate is given. The means of these estimated standard errors (Est-SD) were also calculated and listed in Appendix C.

LMS, LISREL-ML, and LISREL-WLSA are iterative procedures, and it occasionally happens that the iteration does not converge. Analyzing the 500 data sets of a simulated data package, the number of converged solutions ranged from 498 to 500 in LMS (maximum 50 iterations), from 442 to 492 in LISREL-ML (maximum 150 iterations), and from 471 to 500 in LISREL-WLSA (maximum 150 iterations). The data sets involving nonconvergent solutions were removed from the simulation study for each method individually, and the number of remaining valid data sets is indicated in Appendix C.

Because the goal of the simulation study is to detect the methods' advantages and drawbacks for the analysis of interaction models, the parameters of primary interest are the interaction effect γ_3 and the variance ψ_{11} of the disturbance term ζ. For several parts of the simulation study, these parameters were revealed to be the most critical ones.

Bias of Parameter Estimates and Standard Error Estimates. A parameter estimator is unbiased when its expectation value equals the true parameter value, and the mean of 500 estimates from 500 data sets is regarded to be a good approximation of the expectation value. As the means of the parameter estimates indicate (see Appendix C), there is no sign of substantial bias in the estimates of LMS, LISREL-ML, and 2SLS (e.g., see Table 10.3 for bias of $\hat{\gamma}_3$).

Only the LISREL-WLSA method provides some biased estimates when sample size is low ($N = 200$) and γ_3 is medium or high (Model B and Model C). This confirms Yang Jonsson's (1997) statement that LISREL-WLSA should be used only with large samples.

Besides unbiased point estimators, the researcher is interested in estimating correct standard errors to establish confidence intervals for the estimates. Figure 10.2 shows the true standard errors calculated from the Monte Carlo simulation study (MC-SD) in relation to the mean of estimated standard errors (Est-SD). The values are given for the estimator $\hat{\gamma}_3$, Model B ($\gamma_3 = 0.7$) and medium sample size ($N = 400$).

LMS slightly overestimates the true standard error MC-SD by the estimated standard error (Est-SD). We define the relative bias of an estimator by

$$\text{Rel-Bias}(\hat{\theta}) = \left(\frac{\text{Est-SD}(\hat{\theta})}{\text{MC-SD}(\hat{\theta})} - 1 \right) \times 100\%.$$

The relative bias of estimated standard errors in Fig. 10.2 are +8.5% in LMS, −30% in LISREL-ML, −31% in LISREL-WLSA, and −13% in 2SLS. For

TABLE 10.3
Means of Parameter Estimates $\hat{\gamma}_3$ in the Simulation Study

		Interaction Effect		
Method	*Sample Size*	$\gamma_3 = 0.30$ *(Model A)*	$\gamma_3 = 0.70$ *(Model B)*	$\gamma_3 = 1.50$ *(Model C)*
LMS	$N = 200$	0.298	0.692	1.474
LISREL-ML	$N = 200$	0.309	0.714	1.568
LISREL-WLSA	$N = 200$	0.268	0.610	1.302
2SLS	$N = 200$	0.254	0.662	1.450
LMS	$N = 400$	0.306	0.698	1.481
LISREL-ML	$N = 400$	0.307	0.708	1.535
LISREL-WLSA	$N = 400$	0.285	0.654	1.399
2SLS	$N = 400$	0.312	0.691	1.516
LMS	$N = 800$	0.303	0.707	1.514
LISREL-ML	$N = 800$	0.305	0.701	1.515
LISREL-WLSA	$N = 800$	0.292	0.672	1.437
2SLS	$N = 800$	0.305	0.689	1.482

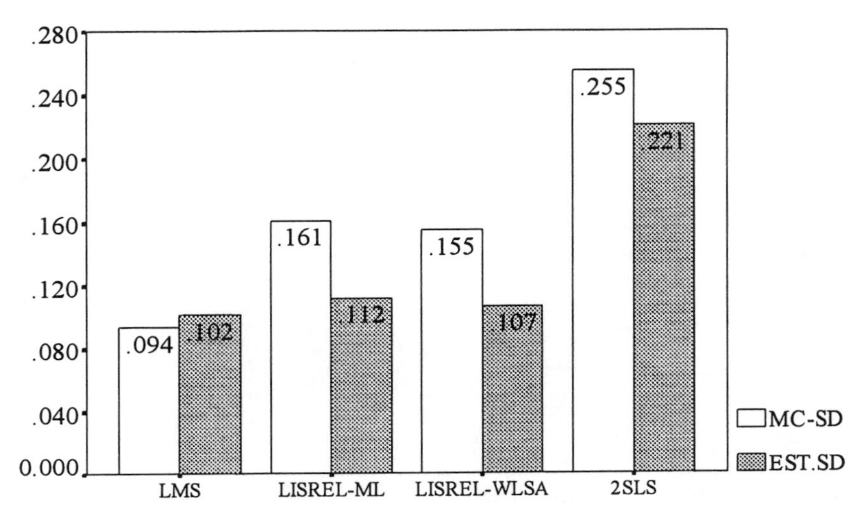

FIG. 10.2. Model B ($\gamma_3 = 0.7$, $N = 400$). Bias of estimated standard errors for $\hat{\gamma}_3$: MC-SD versus Est-SD.

the other parameters, the relative bias of estimated standard errors remains very low in LMS for all models and sample sizes. Thus, the estimated standard errors provided by LMS can be used effectively for confidence intervals about the parameter estimates (see Fig. 10.3).

In Fig. 10.2, the standard errors estimated by LISREL-ML and LISREL-WLSA substantially underestimate the true standard errors (MC-SD) of the estimators of γ_3, so hypothesis testing using estimated standard errors for confidence intervals can be fallible. Asymptotically correct standard errors can be calculated for LISREL-ML by using a correction formula given in Jöreskog and Yang (1997), but the correction still provides biased estimates for sample size $N = 400$ (Klein et al., 1997). For example, the corrected estimated standard error for $\hat{\gamma}_3$ in Model B equals 0.219, now overestimating MC-SD($\hat{\gamma}_3$) = 0.161, with a relative bias of 36%.

The relative underestimation of MC-SD by (uncorrected) Est-SD in LISREL-ML increases substantially with the size of interaction effect γ_3, whereas in LISREL-WLSA, the relative underestimation of MC-SD by Est-SD varies mainly with the sample size N. These results can easily be confirmed by calculating the relative bias for the values given in Appendix C, and matching the performance expected for these methods: Because LISREL-ML assumes normality for the indicators, the robustness decreases with non-normality caused by high interaction; on the other hand, LISREL-WLSA is expected to become problematic when sample size decreases, whereas its performance should not depend on the degree of non-normality as much.

Although the standard errors in 2SLS are clearly higher than in the other methods, the relative bias of 2SLS is lower than in LISREL, but higher than

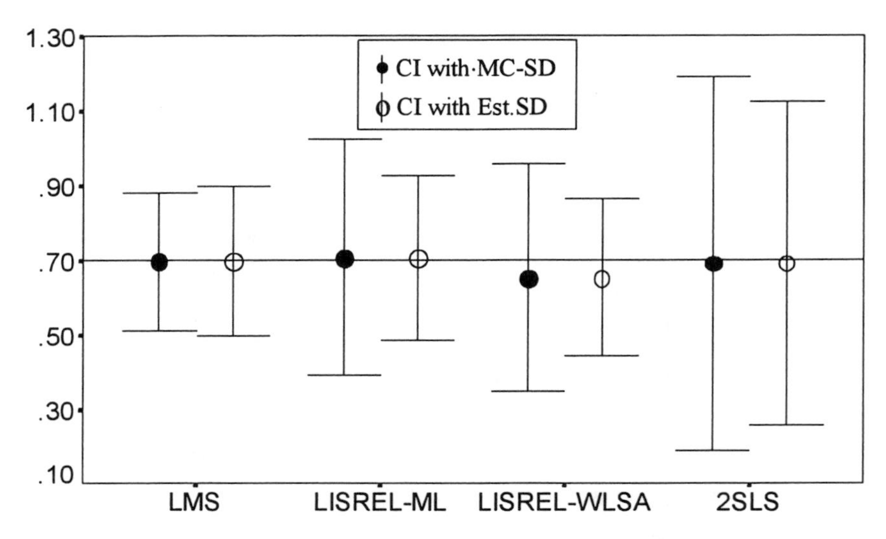

FIG. 10.3. Model B ($\gamma_3 = 0.7$, $N = 400$). Confidence intervals for $\hat{\gamma}_3$: $[M(\hat{\gamma}_3) \pm 1.96 \cdot \text{MC-SD}(\hat{\gamma}_3)]$, $[M(\hat{\gamma}_3) \pm 1.96 \cdot \text{Est-SD}(\hat{\gamma}_3)]$.

in LMS. In general, this tendency can be confirmed for the other models and sample sizes as well (Appendix C).

Figure 10.3 illustrates the 5% confidence intervals calculated from MC-SD standard error versus the intervals calculated from Est-SD standard error. A consequence of underestimating the confidence limits for $\hat{\gamma}_3$ in a single analysis is that the hypothesis H_1: $\gamma_3 \neq 0$ cannot be tested. With LMS and 2SLS, this test can be carried out. The correct estimation of the true confidence intervals allows an inferential interpretation of the point estimate $\hat{\gamma}_3$.

Efficiency of Parameter Estimates. The efficiency of a parameter estimator corresponds to its standard error, and a low standard error indicates high efficiency. Thus, the standard error of an estimator is a measure of its precision, and the lower the standard error (i.e., the higher the efficiency), the higher is the power or detection capability (i.e., the lower is Type II error) of the estimator.

The estimates of all parameters are clearly most efficient in LMS under all investigated conditions (see Appendix C). In order to compare the efficiency of different methods, the relative efficiency (Rinne, 1995) can be calculated. For example, the relative efficiency of LMS versus LISREL-ML for the estimator $\hat{\theta}$ is calculated by

$$\text{Rel-Eff}(\hat{\theta}) = \left(\frac{\text{MC-SD}(\hat{\theta})_{\text{LMS}}}{\text{MC-SD}(\hat{\theta})_{\text{LISREL-ML}}} \right)^2 \times 100\%.$$

TABLE 10.4
Relative Efficiency of LMS Parameter Estimates $\hat{\gamma}_3$ and $\hat{\psi}_{11}$ in Model B ($\gamma_3 = 0.70$) in Relation to LISREL-ML, LISREL-WLSA, and 2SLS

Relative Efficiency	*N = 200*		*N = 400*		*N = 800*	
	$\hat{\gamma}_3$	$\hat{\psi}_{11}$	$\hat{\gamma}_3$	$\hat{\psi}_{11}$	$\hat{\gamma}_3$	$\hat{\psi}_{11}$
LISREL-ML	29.5%	19.9%	34.1%	25.0%	34.3%	19.8%
LISREL-WLSA	36.7%	4.9%	36.8%	7.1%	39.1%	7.1%
2SLS	7.9%	—	13.6%	—	15.3%	—

The relative efficiency values of LMS versus the other methods for the estimators $\hat{\gamma}_3$ and $\hat{\psi}_{11}$ are listed in Table 10.4. As the relative efficiency percentage of parameter estimates $\hat{\gamma}_3$ and $\hat{\psi}_{11}$ for LMS versus LISREL-ML, LISREL-WLSA, and 2SLS is clearly smaller than 100%, this indicates that LMS estimators are definitely more efficient than those of the other estimation methods. LISREL-WLSA and LISREL-ML estimators are more efficient than 2SLS estimators, but differences exist between the two LISREL approaches: Whereas γ_3 is estimated more efficiently by LISREL-WLSA, ψ_{11} is estimated more efficiently by LISREL-ML.

Relative efficiency is closely related to the sample size required by a method for detecting an interaction effect. Hence, LMS requires much smaller sample sizes for the detection of an interaction effect than the other approaches.

DISCUSSION

Over the past decade, the use of structural equation modeling has become increasingly common in the social and behavioral sciences. However, problems with the implementation of structural equation models can occur when the distributions of the observed variables depart substantially from multivariate normality. This is the case for latent interaction models, because they involve nonlinear terms of latent predictor variables in the structural equation. The consequences of nonlinear terms are non-normal distributions of the latent dependent variables, even if the latent predictors follow normal distributions. Moreover, products of normally distributed observed variables as indicators for latent product terms must be formed in some methods, increasing the number of non-normal variates in the model.

The LMS estimation procedure takes the distributional characteristics of non-normally distributed variables explicitly into account. By investigating the density functions occurring in latent interaction models, it gives both theoretical insight into the stochastic structure of latent interaction and implements an iterative ML estimation of the parameters. The simulation study for the elementary interaction model shows that the parameter estimates are unbiased. LMS outperforms the other methods with regard to efficiency, even for small or medium sample sizes. Just as important, the estimation of standard errors in LMS has no substantial bias that supports hypothesis testing of interaction effects.

The LISREL-WLSA method is asymptotically distribution free, so there is no violation of distributional assumptions when applied to interaction models. It provides consistent parameter estimates, but despite these theoretical properties, the application of this method is limited in practice. The simulation study shows that for small ($N = 200$) or medium ($N = 400$) sample sizes, the efficiency of the LISREL-WLSA estimates is very low compared to LMS (see also Jöreskog & Yang, 1996). In addition to that, the standard errors are often grossly underestimated in LISREL-WLSA applications for small and medium sample sizes, and they cannot be interpreted easily for inferential statistics and confidence intervals. The LISREL-ML approach provides consistent and unbiased parameter estimates. There is no general trend whether LISREL-ML or LISREL-WLSA is more efficient. Although the implementation of interaction models in LISREL-ML violates its assumption of normally distributed indicator variables, it can be used in cases where the interaction effect is not too high and sample size is not too small. Both the standard errors given in LISREL-ML outputs and the corrected standard errors (Jöreskog & Yang, 1997) should not be uncritically interpreted for exact inferential statistics. The relative bias of the former rises with increasing interaction effect because of violation of normality assumption; the latter is biased when sample size is small or medium. After all, LISREL-ML and LISREL-WLSA could be valuable tools used for general exploratory purposes or preliminary data analysis, but correct testing of hypotheses and precise parameter estimation cannot be expected for these methods in all cases, especially when sample size is small.

One favorable feature of the 2SLS approach proposed by Bollen (1995, 1996) and Bollen and Paxton (chap. 6) is the easy implementation. The simulation study confirms relatively low bias for standard error estimates, which allows inferential statistics with acceptable Type I error. Still, the disadvantage of 2SLS lies in its low power and low efficiency relative to the other methods examined in the study. In practice, the application of 2SLS requires large sample sizes in order to provide estimates of satisfying precision for interaction models.

The results of the simulation study are quite promising for LMS, which performs well under the models investigated and conditions studied. The new approach, tailored for interaction models, includes some computationally intensive algorithms not yet available in standard statistical software for structural equation models. Further research should concentrate on the optimization of the algorithms, the relaxation of normality assumption for the x-variables, and the generalization of the method for categorized data.

ACKNOWLEDGMENTS

This research has been supported by the Deutsche Forschungsgemeinschaft (DFG), Germany, Grants Mo 474/3-1 and Mo 474/3-2.

REFERENCES

Abramowitz, M., & Stegun, I. A. (1971). *Handbook of mathematical functions*. New York: Dover.

Bollen, K. A. (1995). Structural equation models that are non-linear in latent variables. In P. V. Marsden (Ed.), *Sociological methodology 1995* (Vol. 25). Washington, DC: American Sociological Association.

Bollen, K. A. (1996). An alternative two stage least squares (2SLS) estimator for latent variable equations. *Psychometrika, 61,* 109–121.

Dempster, A. P., Laird, N. M., & Rubin, D. B. (1977). Maximum likelihood from incomplete data via the EM algorithm. *Journal of the Royal Statistical Society, Ser. B, 39,* 1–38.

Isaacson, E., & Keller, H. B. (1966). *Analysis of numerical methods*. New York: Wiley.

Jöreskog, K. G., & Sörbom, D. (1989). *LISREL 7: User's reference guide*. Mooresville, IN: Scientific Software.

Jöreskog, K. G., & Sörbom, D. (1996a). *LISREL 8 User's reference guide*. Chicago: Scientific Software.

Jöreskog, K. G., & Sörbom, D. (1996b). *PRELIS 2 User's guide*. Chicago: Scientific Software.

Jöreskog, K. G., & Yang, F. (1996). Non-linear structural equation models: The Kenny–Judd model with interaction effects. In G. Marcoulides & R. Schumacker (Eds.), *Advanced structural equation modeling* (pp. 57–87). Mahwah, NJ: Lawrence Erlbaum Associates.

Jöreskog, K. G., & Yang, F. (1997). Estimation of interaction models using the augmented moment matrix: Comparison of asymptotic standard errors. In W. Bandilla & F. Faulbaum (Eds.), *SoftStat '97* (Advances in statistical software 6, pp. 467–478). Stuttgart: Lucius & Lucius.

Kenny, D. A., & Judd, C. M. (1984). Estimating the non-linear and interactive effects of latent variables. *Psychological Bulletin, 99,* 422–431.

Klein, A., & Moosbrugger, H. (1997). *Maximum likelihood estimations of latent interaction effects with the LMS-method.* Manuscript submitted.

Klein, A., Moosbrugger, H., Schermelleh-Engel, K., & Frank, D. (1997). A new approach to the estimation of latent interaction effects in structural equation models. In W. Bandilla & F. Faulbaum (Eds.), *SoftStat '97* (Advances in statistical software 6, pp. 479–486). Stuttgart: Lucius & Lucius.

Moosbrugger, H., Schermelleh-Engel, K., & Klein, A. (1997). Methodological problems of estimating latent interaction effects. *Methods of Psychological Research Online, 2*, 95–111.

Redner, R. A., & Walker, H. F. (1984). Mixture densities, maximum likelihood and the EM algorithm. *SIAM Review, 26,* 195–239.

Rinne, H. (1995). *Taschenbuch der Statistik* [Handbook of statistics]. Frankfurt am Main: Harri Deutsch.

Schwarz, H. R. (1993). *Numerische Mathematik* [Numerical mathematics]. Stuttgart: Teubner.

Yang Jonsson, F. (1997). *Non-linear structural equation models: Simulation studies of the Kenny–Judd model.* Uppsala, Sweden: University of Uppsala.

APPENDIX A
LISREL 8 Input File for ML

```
!Elementary Interaction Model for ML With Four Product Variables,
 TX = 0
DA NI=9 NO=200 RP=500
ME=ML207.ME
CM=ML207.CM
LA
Y1 X1 X2 X3 X4 X1X3 X1X4 X2X3 X2X4
MO NX=8 NY=1 NK=3 NE=1 TD=SY PS=FU,FR PH=SY GA=FU LX=FU LY=FI KA=FI
 AL=FR
LK
KSI1 KSI2 KSI1KSI2
LE
ETA1
PA LX
1 0 0
1 0 0
0 1 0
0 1 0
0 0 1
0 0 1
0 0 1
0 0 1
FI LX(1,1) LX(3,2) LX(5,3)
VA 1 LX(1,1) LX(3,2) LX(5,3) LY(1,1)
PA TD
1
0 1
0 0 1
0 0 0 1
0 0 0 0 1
0 0 0 0 1 1
0 0 0 0 1 0 1
0 0 0 0 0 1 1 1
FI TE(1,1)
PA PH
1
1 1
0 0 1
PA GA
1 1 1
```

```
PA KA
0 0 1
CO PH(3,3)=PH(1,1)*PH(2,2)+PH(2,1)**2
CO KA(3)=PH(2,1)
EQ LX(6,3)=LX(4,2)
EQ LX(7,3)=LX(2,1)
CO LX(8,3)=LX(2,1)*LX(4,2)
CO TD(5,5)=PH(1,1)*TD(3,3)+PH(2,2)*TD(1,1)+TD(1,1)*TD(3,3)
CO TD(6,6)=PH(1,1)*TD(4,4)+LX(4,2)**2*PH(2,2)*TD(1,1)+TD(1,1)
 *TD(4,4)
CO TD(7,7)=LX(2,1)**2*PH(1,1)*TD(3,3)+PH(2,2)*TD(2,2)+TD(2,2)
 *TD(3,3)
CO TD(8,8)=LX(2,1)**2*PH(1,1)*TD(4,4)+LX(4,2)**2*PH(2,2)*TD(2,2)
 +TD(2,2)*TD(4,4)
CO TD(6,5)=LX(4,2)*PH(2,2)*TD(1,1)
CO TD(7,5)=LX(2,1)*PH(1,1)*TD(3,3)
CO TD(8,6)=LX(2,1)*PH(1,1)*TD(4,4)
CO TD(8,7)=LX(4,2)*PH(2,2)*TD(2,2)
OU AD=OFF IT=150 ND=3 EP=0.0001 PV=ML207.PV SV=ML207.SV XM
```

APPENDIX B
LISREL 8 Input File for WLSA

```
!Elementary Interaction Model for WLSA With Four Product Variables,
 TX = 0
DA NI=10 NO=200 RP=500
LA
Y X1 X2 X3 X4 X1X3 X1X4 X2X3 X2X4 CONST
CM=WLSA207.AM
WM=WLSA207.WMA
MO NY=9 NE=4 NX=1 GA=FI TE=SY PS=SY,FI FI
FR LY(1,1) LY(1,2) LY(1,3) LY(1,4)
FR LY(3,1) LY(5,2)
VA 1 LY(2,1) LY(4,2) LY(6,3)
VA 1 GA(4)
CO GA(3)=PS(2,1)
CO LY(7,3)=LY(5,2)
CO LY(8,3)=LY(3,1)
CO LY(9,3)=LY(3,1)*LY(5,2)
FR PS(1,1)-PS(2,2)
FI PS(3,1) PS(3,2)
CO PS(3,3)=PS(1,1)*PS(2,2)+PS(2,1)**2
CO TE(6,6)=PS(1,1)*TE(4,4)+PS(2,2)*TE(2,2)+TE(2,2)*TE(4,4)
CO TE(7,6)=LY(5,2)*PS(2,2)*TE(2,2)
CO TE(7,7)=PS(1,1)*TE(5,5)+LY(5,2)**2*PS(2,2)*TE(2,2)+TE(2,2)
```

```
 *TE(5,5)
CO TE(8,6)=LY(3,1)*PS(1,1)*TE(4,4)
CO TE(8,8)=LY(3,1)**2*PS(1,1)*TE(4,4)+PS(2,2)*TE(3,3)+TE(3,3)
 *TE(4,4)
CO TE(9,7)=LY(3,1)*PS(1,1)*TE(5,5)
CO TE(9,8)=LY(5,2)*PS(2,2)*TE(3,3)
CO TE(9,9)=LY(3,1)**2*PS(1,1)*TE(5,5)+LY(5,2)**2*PS(2,2)*TE(3,3)
 +TE(3,3)*TE(5,5)
ST .5 ALL
ST 1 LY(1,4)
OU SO NS AD=OFF DF=-9 IT=150 PV=WLSA207.PV SV=WLSA207.SV XM XI
```

APPENDIX C
Results of the Monte Carlo Studies

TABLE 10.C1
LMS Results of Monte Carlo Studies With $\gamma_3 = 0.3$, $N = 200$, 400, and 800: Mean (M) and Standard Error (MC-SD) of Parameter Estimates, and Estimated Standard Error (Est-SD)

		LMS: $\gamma_3 = 0.3$ (Model A)								
		$N = 200$ (Valid Data Sets = 498)			*$N = 400$ (Valid Data Sets = 500)*			*$N = 800$ (Valid Data Sets = 500)*		
Parameter	*True Value*	*M*	*MC-SD*	*Est-SD*	*M*	*MC-SD*	*Est-SD*	*M*	*MC-SD*	*Est-SD*
γ_1	0.20	0.203	0.103	0.101	0.192	0.063	0.065	0.188	0.044	0.047
γ_2	0.40	0.409	0.089	0.080	0.412	0.061	0.059	0.410	0.042	0.042
γ_3	0.30	0.298	0.099	0.099	0.306	0.071	0.068	0.303	0.049	0.050
α	1.00	1.002	0.041	0.040	0.999	0.030	0.029	1.000	0.020	0.021
ψ_{11}	0.20	0.186	0.028	0.027	0.194	0.020	0.020	0.198	0.014	0.014
λ_{21}	0.60	0.605	0.157	0.162	0.600	0.107	0.114	0.594	0.069	0.076
λ_{42}	0.70	0.711	0.117	0.111	0.711	0.079	0.080	0.715	0.058	0.057
ϕ_{11}	0.49	0.488	0.115	0.121	0.504	0.088	0.102	0.503	0.056	0.059
ϕ_{21}	0.2352	0.226	0.066	0.062	0.238	0.047	0.045	0.235	0.033	0.032
ϕ_{22}	0.64	0.621	0.121	0.110	0.631	0.090	0.073	0.629	0.062	0.053
θ_{11}	0.51	0.480	0.093	0.099	0.500	0.075	0.083	0.498	0.042	0.043
θ_{22}	0.64	0.614	0.077	0.079	0.636	0.051	0.056	0.641	0.041	0.041
θ_{33}	0.36	0.352	0.093	0.097	0.366	0.065	0.065	0.369	0.048	0.047
θ_{44}	0.51	0.484	0.061	0.066	0.505	0.049	0.048	0.505	0.035	0.034

TABLE 10.C2

LMS Results of Monte Carlo Studies With $\gamma_3 = 0.7$, $N = 200$, 400, and 800: Mean (M) and Standard Error (MC-SD) of Parameter Estimates, and Estimated Standard Error (Est-SD)

		LMS: $\gamma_3 = 0.7$ *(Model B)*								
		$N = 200$ *(Valid Data Sets = 500)*			$N = 400$ *(Valid Data Sets = 500)*			$N = 800$ *(Valid Data Sets = 500)*		
Parameter	*True Value*	*M*	*MC-SD*	*Est-SD*	*M*	*MC-SD*	*Est-SD*	*M*	*MC-SD*	*Est-SD*
γ_1	0.20	0.196	0.099	0.093	0.196	0.064	0.065	0.201	0.047	0.048
γ_2	0.40	0.404	0.088	0.086	0.411	0.061	0.061	0.403	0.039	0.043
γ_3	0.70	0.692	0.132	0.136	0.698	0.094	0.102	0.707	0.065	0.073
α	1.00	0.999	0.047	0.046	1.002	0.032	0.033	0.998	0.024	0.023
ψ_{11}	0.20	0.192	0.033	0.033	0.195	0.025	0.024	0.193	0.016	0.017
λ_{21}	0.60	0.598	0.127	0.132	0.593	0.092	0.099	0.601	0.067	0.073
λ_{42}	0.70	0.715	0.115	0.117	0.703	0.077	0.078	0.709	0.055	0.054
ϕ_{11}	0.49	0.496	0.108	0.114	0.505	0.077	0.081	0.471	0.052	0.058
ϕ_{21}	0.2352	0.234	0.062	0.060	0.235	0.044	0.040	0.231	0.030	0.028
ϕ_{22}	0.64	0.626	0.118	0.105	0.637	0.087	0.086	0.615	0.056	0.054
θ_{11}	0.51	0.478	0.082	0.095	0.498	0.059	0.061	0.498	0.040	0.042
θ_{22}	0.64	0.615	0.072	0.075	0.633	0.050	0.054	0.623	0.038	0.038
θ_{33}	0.36	0.347	0.089	0.084	0.361	0.060	0.057	0.357	0.042	0.041
θ_{44}	0.51	0.493	0.068	0.063	0.506	0.046	0.047	0.493	0.033	0.032

TABLE 10.C3

LMS Results of Monte Carlo Studies With $\gamma_3 = 1.5$, $N = 200$, 400, and 800: Mean (M) and Standard Error (MC-SD) of Parameter Estimates, and Estimated Standard Error (Est-SD)

		LMS: $\gamma_3 = 1.5$ (Model C)								
		N = 200 (Valid Data Sets = 500)			*N = 400 (Valid Data Sets = 500)*			*N = 800 (Valid Data Sets = 500)*		
Parameter	*True Value*	*M*	*MC-SD*	*Est-SD*	*M*	*MC-SD*	*Est-SD*	*M*	*MC-SD*	*Est-SD*
γ_1	0.20	0.196	0.117	0.116	0.196	0.084	0.084	0.202	0.056	0.059
γ_2	0.40	0.396	0.104	0.107	0.399	0.076	0.075	0.402	0.052	0.053
γ_3	1.50	1.474	0.236	0.230	1.481	0.161	0.158	1.514	0.110	0.115
α	1.00	1.003	0.058	0.058	1.004	0.042	0.042	0.999	0.030	0.030
ψ_{11}	0.20	0.198	0.053	0.051	0.203	0.038	0.037	0.197	0.027	0.026
λ_{21}	0.60	0.601	0.118	0.114	0.606	0.086	0.086	0.613	0.060	0.063
λ_{42}	0.70	0.702	0.096	0.098	0.699	0.074	0.068	0.702	0.045	0.049
ϕ_{11}	0.49	0.485	0.097	0.105	0.493	0.071	0.077	0.468	0.051	0.053
ϕ_{21}	0.2352	0.232	0.058	0.050	0.238	0.041	0.037	0.227	0.030	0.025
ϕ_{22}	0.64	0.636	0.107	0.095	0.648	0.083	0.079	0.624	0.057	0.053
θ_{11}	0.51	0.485	0.076	0.079	0.502	0.053	0.059	0.501	0.040	0.041
θ_{22}	0.64	0.615	0.070	0.069	0.632	0.050	0.051	0.620	0.035	0.036
θ_{33}	0.36	0.344	0.070	0.068	0.353	0.050	0.049	0.350	0.034	0.035
θ_{44}	0.51	0.492	0.058	0.060	0.510	0.047	0.045	0.494	0.030	0.031

TABLE 10.C4

LISREL-ML Results of Monte Carlo Studies With $\gamma_3 = 0.3$, $N = 200$, 400, and 800: Mean (M) and Standard Error (MC-SD) of Parameter Estimates, and Estimated Standard Error (Est-SD)

		LISREL-ML: $\gamma_3 = 0.3$ (Model A)								
		$N = 200$ (Valid Data Sets = 450)			*$N = 400$ (Valid Data Sets = 473)*			*$N = 800$ (Valid Data Sets = 483)*		
Parameter	*True Value*	*M*	*MC-SD*	*Est-SD*	*M*	*MC-SD*	*Est-SD*	*M*	*MC-SD*	*Est-SD*
γ_1	0.20	0.202	0.199	0.123	0.202	0.080	0.070	0.202	0.056	0.049
γ_2	0.40	0.402	0.113	0.091	0.402	0.069	0.059	0.400	0.047	0.040
γ_3	0.30	0.309	0.153	0.114	0.307	0.098	0.078	0.305	0.067	0.055
α	1.00	1.000	0.048	0.043	1.000	0.034	0.031	1.000	0.022	0.022
ψ_{11}	0.20	0.192	0.040	0.034	0.195	0.025	0.023	0.198	0.018	0.016
λ_{12}	0.60	0.632	0.220	0.143	0.612	0.148	0.097	0.608	0.099	0.068
λ_{42}	0.70	0.713	0.128	0.091	0.699	0.093	0.064	0.698	0.064	0.045
ϕ_{11}	0.49	0.549	0.455	0.265	0.509	0.142	0.095	0.495	0.094	0.064
ϕ_{21}	0.2352	0.238	0.073	0.043	0.237	0.049	0.031	0.234	0.034	0.022
ϕ_{22}	0.64	0.651	0.140	0.100	0.644	0.106	0.070	0.643	0.072	0.050
θ_{11}	0.51	0.456	0.446	0.264	0.493	0.133	0.094	0.503	0.081	0.063
θ_{22}	0.64	0.633	0.103	0.063	0.634	0.063	0.043	0.640	0.049	0.031
θ_{33}	0.36	0.346	0.103	0.087	0.354	0.078	0.062	0.357	0.056	0.043
θ_{44}	0.51	0.499	0.077	0.053	0.509	0.055	0.037	0.510	0.038	0.027

TABLE 10.C5

LISREL-ML Results of Monte Carlo Studies With $\gamma_3 = 0.7$, $N = 200$, 400, and 800: Mean (M) and Standard Error (MC-SD) of Parameter Estimates, and Estimated Standard Error (Est-SD)

		LISREL-ML: $\gamma_3 = 0.7$ (Model B)								
		$N = 200$ (Valid Data Sets = 458)			*$N = 400$ (Valid Data Sets = 470)*			*$N = 800$ (Valid Data Sets = 485)*		
Parameter	*True Value*	*M*	*MC-SD*	*Est-SD*	*M*	*MC-SD*	*Est-SD*	*M*	*MC-SD*	*Est-SD*
γ_1	0.20	0.200	0.136	0.118	0.211	0.089	0.076	0.197	0.060	0.052
γ_2	0.40	0.404	0.124	0.095	0.400	0.076	0.062	0.402	0.051	0.044
γ_3	0.70	0.714	0.243	0.162	0.708	0.161	0.112	0.701	0.111	0.078
α	1.00	0.994	0.066	0.054	1.000	0.042	0.038	1.000	0.031	0.027
ψ_{11}	0.20	0.188	0.074	0.057	0.193	0.050	0.039	0.197	0.036	0.027
λ_{12}	0.60	0.620	0.191	0.113	0.609	0.121	0.079	0.602	0.086	0.055
λ_{42}	0.70	0.703	0.131	0.084	0.697	0.092	0.060	0.697	0.063	0.042
ϕ_{11}	0.49	0.483	0.755	0.484	0.500	0.129	0.078	0.499	0.083	0.054
ϕ_{21}	0.2352	0.244	0.071	0.043	0.235	0.049	0.031	0.235	0.036	0.022
ϕ_{22}	0.64	0.652	0.149	0.096	0.653	0.110	0.068	0.644	0.076	0.047
θ_{11}	0.51	0.519	0.751	0.481	0.500	0.109	0.075	0.506	0.073	0.052
θ_{22}	0.64	0.632	0.090	0.056	0.632	0.060	0.039	0.636	0.044	0.028
θ_{33}	0.36	0.347	0.114	0.082	0.350	0.077	0.058	0.357	0.053	0.040
θ_{44}	0.51	0.502	0.077	0.051	0.507	0.052	0.036	0.510	0.036	0.025

TABLE 10.C6

LISREL-ML Results of Monte Carlo Studies With $\gamma_3 = 1.5$, $N = 200$, 400, and 800: Mean (M) and Standard Error (MC-SD) of Parameter Estimates, and Estimated Standard Error (Est-SD)

		LISREL-ML: $\gamma_3 = 1.5$ *(Model C)*								
		$N = 200$ *(Valid Data Sets = 445)*			$N = 400$ *(Valid Data Sets = 442)*			$N = 800$ *(Valid Data Sets = 492)*		
Parameter	*True Value*	*M*	*MC-SD*	*Est-SD*	*M*	*MC-SD*	*Est-SD*	*M*	*MC-SD*	*Est-SD*
γ_1	0.20	0.233	0.242	0.174	0.205	0.127	0.106	0.204	0.086	0.072
γ_2	0.40	0.387	0.169	0.128	0.402	0.111	0.084	0.400	0.078	0.058
γ_3	1.50	1.568	0.476	0.287	1.535	0.309	0.190	1.515	0.212	0.132
α	1.00	0.986	0.111	0.085	0.996	0.070	0.059	1.002	0.054	0.041
ψ_{11}	0.20	0.151	0.239	0.159	0.182	0.146	0.105	0.189	0.111	0.075
λ_{12}	0.60	0.622	0.159	0.094	0.618	0.116	0.066	0.597	0.076	0.046
λ_{42}	0.70	0.699	0.129	0.079	0.705	0.089	0.055	0.708	0.062	0.039
ϕ_{11}	0.49	0.501	0.224	0.109	0.484	0.112	0.064	0.495	0.084	0.047
ϕ_{21}	0.2352	0.240	0.071	0.042	0.236	0.050	0.030	0.235	0.033	0.021
ϕ_{22}	0.64	0.646	0.147	0.091	0.641	0.110	0.063	0.636	0.072	0.044
θ_{11}	0.51	0.486	0.210	0.103	0.504	0.089	0.060	0.506	0.069	0.043
θ_{22}	0.64	0.635	0.087	0.052	0.630	0.057	0.036	0.640	0.043	0.026
θ_{33}	0.36	0.356	0.109	0.076	0.354	0.072	0.052	0.361	0.051	0.036
θ_{44}	0.51	0.509	0.076	0.048	0.506	0.053	0.034	0.506	0.036	0.024

TABLE 10.C7

LISREL-WLSA Results of Monte Carlo Studies With $\gamma_3 = 0.3$, $N = 200$, 400, and 800: Mean (M) and Standard Error (MC-SD) of Parameter Estimates, and Estimated Standard Error (Est-SD)

		LISREL-WLSA: $\gamma_3 = 0.3$ *(Model A)*								
		$N = 200$ *(Valid Data Sets = 471)*			$N = 400$ *(Valid Data Sets = 499)*			$N = 800$ *(Valid Data Sets = 500)*		
Parameter	*True Value*	*M*	*MC-SD*	*Est-SD*	*M*	*MC-SD*	*Est-SD*	*M*	*MC-SD*	*Est-SD*
γ_1	0.20	0.200	0.149	0.100	0.198	0.087	0.068	0.198	0.056	0.049
γ_2	0.40	0.386	0.106	0.081	0.395	0.070	0.059	0.396	0.048	0.042
γ_3	0.30	0.268	0.155	0.084	0.285	0.095	0.068	0.292	0.066	0.053
α	1.00	1.005	0.048	0.040	1.003	0.036	0.029	1.001	0.023	0.022
ψ_{11}	0.20	0.188	0.094	0.092	0.198	0.066	0.069	0.202	0.039	0.050
λ_{21}	0.60	0.678	0.251	0.126	0.643	0.171	0.102	0.620	0.102	0.077
λ_{42}	0.70	0.744	0.163	0.086	0.717	0.097	0.068	0.708	0.066	0.051
ϕ_{11}	0.49	0.567	0.258	0.147	0.524	0.159	0.100	0.504	0.092	0.073
ϕ_{21}	0.2352	0.274	0.090	0.039	0.256	0.059	0.033	0.246	0.039	0.025
ϕ_{22}	0.64	0.652	0.160	0.090	0.643	0.105	0.073	0.644	0.074	0.056
θ_{11}	0.51	0.381	0.253	0.166	0.433	0.159	0.115	0.475	0.087	0.084
θ_{22}	0.64	0.538	0.155	0.088	0.578	0.092	0.072	0.605	0.061	0.055
θ_{33}	0.36	0.234	0.140	0.103	0.296	0.087	0.083	0.329	0.063	0.064
θ_{44}	0.51	0.374	0.126	0.081	0.436	0.082	0.068	0.470	0.048	0.053

TABLE 10.C8

LISREL-WLSA Results of Monte Carlo Studies With $\gamma_3 = 0.7$, $N = 200$, 400, and 800: Mean (M) and Standard Error (MC-SD) of Parameter Estimates, and Estimated Standard Error (Est-SD)

		LISREL-WLSA: $\gamma_3 = 0.7$ *(Model B)*								
		$N = 200$ *(Valid Data Sets = 490)*			$N = 400$ *(Valid Data Sets = 500)*			$N = 800$ *(Valid Data Sets = 500)*		
Parameter	*True Value*	*M*	*MC-SD*	*Est-SD*	*M*	*MC-SD*	*Est-SD*	*M*	*MC-SD*	*Est-SD*
γ_1	0.20	0.202	0.157	0.107	0.203	0.091	0.075	0.195	0.062	0.053
γ_2	0.40	0.378	0.131	0.091	0.393	0.079	0.065	0.397	0.051	0.047
γ_3	0.70	0.610	0.218	0.131	0.654	0.155	0.107	0.672	0.104	0.085
α	1.00	1.013	0.065	0.049	1.008	0.044	0.037	1.003	0.030	0.027
ψ_{11}	0.20	0.215	0.149	0.128	0.223	0.094	0.101	0.212	0.060	0.074
λ_{21}	0.60	0.636	0.226	0.107	0.629	0.136	0.085	0.618	0.094	0.067
λ_{42}	0.70	0.744	0.164	0.083	0.712	0.100	0.067	0.706	0.066	0.051
ϕ_{11}	0.49	0.569	0.261	0.161	0.521	0.140	0.086	0.506	0.086	0.065
ϕ_{21}	0.2352	0.268	0.089	0.039	0.255	0.058	0.032	0.247	0.041	0.025
ϕ_{22}	0.64	0.646	0.161	0.091	0.650	0.115	0.074	0.644	0.073	0.056
θ_{11}	0.51	0.366	0.246	0.175	0.452	0.126	0.099	0.475	0.085	0.074
θ_{22}	0.64	0.527	0.137	0.081	0.577	0.091	0.068	0.611	0.059	0.053
θ_{33}	0.36	0.230	0.149	0.101	0.290	0.093	0.081	0.327	0.060	0.062
θ_{44}	0.51	0.381	0.127	0.081	0.443	0.078	0.067	0.474	0.049	0.053

TABLE 10.C9
LISREL-WLSA Results of Monte Carlo Studies With $\gamma_3 = 1.5$, $N = 200$, 400, and 800: Mean (M) and Standard Error (MC-SD) of Parameter Estimates, and Estimated Standard Error (Est-SD)

		LISREL-WLSA: $\gamma_3 = 1.5$ (Model C)								
		N = 200 (Valid Data Sets = 476)			*N = 400 (Valid Data Sets = 500)*			*N = 800 (Valid Data Sets = 500)*		
Parameter	*True Value*	*M*	*MC-SD*	*Est-SD*	*M*	*MC-SD*	*Est-SD*	*M*	*MC-SD*	*Est-SD*
γ_1	0.20	0.219	0.238	0.159	0.198	0.130	0.104	0.195	0.088	0.074
γ_2	0.40	0.360	0.191	0.135	0.394	0.115	0.091	0.399	0.079	0.067
γ_3	1.50	1.302	0.396	0.220	1.399	0.282	0.194	1.437	0.192	0.155
α	1.00	1.024	0.107	0.072	1.010	0.065	0.057	1.009	0.051	0.044
ψ_{11}	0.20	0.285	0.328	0.280	0.249	0.220	0.224	0.235	0.144	0.171
λ_{21}	0.60	0.657	0.200	0.090	0.640	0.135	0.076	0.609	0.085	0.058
λ_{42}	0.70	0.728	0.161	0.075	0.722	0.107	0.063	0.715	0.068	0.049
ϕ_{11}	0.49	0.552	0.238	0.130	0.504	0.124	0.076	0.511	0.091	0.060
ϕ_{21}	0.2352	0.273	0.090	0.037	0.256	0.059	0.031	0.249	0.039	0.024
ϕ_{22}	0.64	0.655	0.167	0.088	0.642	0.114	0.070	0.633	0.073	0.054
θ_{11}	0.51	0.356	0.240	0.141	0.444	0.103	0.086	0.473	0.075	0.068
θ_{22}	0.64	0.515	0.135	0.078	0.576	0.088	0.066	0.614	0.063	0.052
θ_{33}	0.36	0.244	0.145	0.095	0.290	0.089	0.077	0.328	0.058	0.059
θ_{44}	0.51	0.386	0.135	0.079	0.436	0.080	0.067	0.465	0.048	0.051

TABLE 10.C10

2SLS Results of Monte Carlo Studies With $\gamma_3 = 0.3$, $N = 200$, 400, and 800: Mean (M) and Standard Error (MC-SD) of Parameter Estimates, and Estimated Standard Error (Est-SD)

		2SLS: $\gamma_3 = 0.3$ (Model A)								
		$N = 200$ (Valid Data Sets = 500)			$N = 400$ (Valid Data Sets = 500)			$N = 800$ (Valid Data Sets = 500)		
Parameter	*True Value*	*M*	*MC-SD*	*Est-SD*	*M*	*MC-SD*	*Est-SD*	*M*	*MC-SD*	*Est-SD*
γ_1	0.20	0.209	0.202	0.222	0.199	0.123	0.128	0.197	0.080	0.083
γ_2	0.40	0.396	0.130	0.150	0.404	0.087	0.090	0.401	0.059	0.058
γ_3	0.30	0.254	0.372	0.352	0.312	0.216	0.180	0.305	0.116	0.105
α	1.00	1.011	0.085	0.091	0.997	0.057	0.053	0.999	0.034	0.033

TABLE 10.C11

2SLS Results of Monte Carlo Studies with $\gamma_3 = 0.7$, $N = 200$, 400, and 800: Mean (M) and Standard Error (MC-SD) of Parameter Estimates, and Estimated Standard Error (Est-SD)

		2SLS: $\gamma_3 = 0.7$ (Model B)								
		$N = 200$ (Valid Data Sets = 500)			*$N = 400$ (Valid Data Sets = 500)*			*$N = 800$ (Valid Data Sets = 500)*		
Parameter	*True Value*	*M*	*MC-SD*	*Est-SD*	*M*	*MC-SD*	*Est-SD*	*M*	*MC-SD*	*Est-SD*
γ_1	0.20	0.214	0.324	0.303	0.215	0.156	0.165	0.198	0.107	0.110
γ_2	0.40	0.407	0.208	0.200	0.396	0.114	0.115	0.398	0.074	0.077
γ_3	0.70	0.662	0.468	0.435	0.691	0.255	0.221	0.689	0.166	0.143
α	1.00	1.013	0.105	0.114	1.005	0.067	0.066	1.003	0.044	0.043

TABLE 10.C12
2SLS Results of Monte Carlo Studies With $\gamma_3 = 1.5$, $N = 200$, 400, and 800: Mean (M) and Standard Error (MC-SD) of Parameter Estimates, and Estimated Standard Error (Est-SD)

		2SLS: $\gamma_3 = 1.5$ (Model C)								
		N = 200 (Valid Data Sets = 500)			*N = 400 (Valid Data Sets = 500)*			*N = 800 (Valid Data Sets = 500)*		
Parameter	*True Value*	*M*	*MC-SD*	*Est-SD*	*M*	*MC-SD*	*Est-SD*	*M*	*MC-SD*	*Est-SD*
γ_1	0.20	0.175	0.488	0.504	0.206	0.268	0.286	0.213	0.198	0.186
γ_2	0.40	0.393	0.337	0.339	0.393	0.212	0.200	0.399	0.134	0.131
γ_3	1.50	1.450	1.197	0.905	1.516	0.470	0.393	1.482	0.285	0.243
α	1.00	1.013	0.257	0.224	0.999	0.118	0.116	1.006	0.073	0.073

11

Interaction and Nonlinear Modeling: Issues and Approaches

Karl G. Jöreskog
Uppsala University

Kenny and Judd (1984) formulated the first nonlinear structural equation model. As witnessed by the chapters in this book, this stimulated several researchers in the 1990s to consider various procedures for estimating and testing such models. This chapter considers the meaning of interaction modeling in different situations and reviews most of these procedures.

Suppose one wants to estimate the relationship between a dependent variable y and an independent variable x. Interaction modeling usually means that this relationship changes with the values of another variable v. A more general definition is that the distribution of y and x changes with v. The variable v is an interaction variable or moderator variable in the sense that it interacts with or moderates the distribution of y and x or the relationship between y and x.

In general, y and x may consist of several variables and may be either observed variables that are directly measured or latent variables that are unobservable and measured only indirectly via observed indicators containing measurement errors. If they are latent variables, the notations η and ξ will be used instead of y and x.

OBSERVED VARIABLES

Consider first the case where all variables are observed. If y, x, and v are categorical (nominal or ordinal), interaction problems may be investigated by means of log-linear modeling (Agresti, 1990) or graphical modeling (Edwards, 1995). For example, one can test whether the joint distribution of y and x is the same for all values of v or whether y is conditionally independent of v for given x values. If y is categorical and x is a continuous variable, one can use a logistic or a probit regression or other type of generalized linear model of y on x and investigate how this regression varies with different values of v (McCullagh & Nelder, 1989).

If both y and x are continuous variables measured on an interval or a ratio scale, one is usually interested in estimating a linear regression equation of the form

$$y = \alpha + \gamma x + z, \tag{1}$$

where z is an error term assumed to be uncorrelated with x. There are three ways in which this relationship can depend on v:

1. $\alpha = f(v)$,
2. $\gamma = g(v)$,
3. The error variance $Var(z)$ depends on v,

where $f(v)$ and $g(v)$ are some unknown functions of v usually assumed to be linear. One or more of these cases may hold simultaneously.

If v is a discrete variable taking on a few distinct values or if v can be grouped in such a way, all three cases can be handled by the multigroup option in any SEM program. In case 1, the regression equations are assumed to have the same slopes γ but the intercept terms α are allowed to be different in different groups. In case 2, the slopes γ are allowed to be different in different groups whereas the intercepts α and the error variances are assumed to be equal across groups. Case 3 is usually handled under the term *heterogeneous regression*, that is, the regression equation is assumed to be the same across groups but the error variance is allowed to be different in different groups. Two or more of these cases may hold simultaneously.

Next, consider the case where the moderator variable v is continuous. Case 3 does not appear to have been studied in the literature and therefore only cases 1 and 2 are considered in the following.

Case 1 with $f(v) = a + bv$ leads to the linear equation

$$y = a + bv + \gamma x + z. \tag{2}$$

If z is uncorrelated with both x and v, this can be estimated by ordinary least squares and b can be interpreted as an interaction effect: An increase of one unit of v increases the intercept in the regression of y on x by bv y-units.

Case 2 with $g(v) = c + dv$ leads to the nonlinear equation

$$y = \alpha + cx + dxv + z. \tag{3}$$

If z is uncorrelated with both x and v, this can also be estimated by ordinary least squares and d can be interpreted as an interaction effect: An increase of one unit of v increases the slope in the regression of y on x by dv y-units.

If both 1 and 2 hold, then

$$y = a + bv + cx + dxv + z. \tag{4}$$

If z is uncorrelated with both x and v, this can also be estimated by ordinary least squares.

LATENT VARIABLES

Consider the situation where η and ξ are continuous latent variables. We are interested in studying how the relationship

$$\eta = \alpha + \gamma\xi + \zeta, \tag{5}$$

where ζ is uncorrelated with ξ, depends on the interaction variable v. It is assumed that there are p indicators $\mathbf{y}$ of η and q indicators $\mathbf{x}$ of ξ:

$$\mathbf{y}' = (y_1, y_2, \ldots, y_p) \quad \mathbf{x}' = (x_1, x_2, \ldots, x_q)$$

such that (using standard LISREL notation)

$$\mathbf{y} = \boldsymbol{\tau}_y + \boldsymbol{\Lambda}_y\eta + \boldsymbol{\varepsilon}, \tag{6}$$

and

$$\mathbf{x} = \boldsymbol{\tau}_x + \boldsymbol{\Lambda}_x\xi + \boldsymbol{\delta}, \tag{7}$$

where $\boldsymbol{\tau}_y$ and $\boldsymbol{\tau}_x$ are vectors of intercept terms, $\boldsymbol{\Lambda}_y$ and $\boldsymbol{\Lambda}_x$ are regression matrices (or factor loadings), and $\boldsymbol{\varepsilon}$ and $\boldsymbol{\delta}$ are vectors of error terms (errors of measurement or measure-specific components) assumed to be independent of η and ξ, respectively. Cases 1 and 2 of the previous section remain the same: (1) α may depend on v, and/or (2) γ may depend on v.

If v is an observed discrete variable taking on G distinct values, cases 1 and 2 can be investigated by grouping the data into G groups according to the value of v and then view the problem as a multiple-group problem. As pointed out by Rigdon et al. (chap. 1, this volume), this is a simple and straightforward approach that can be used with most SEM programs. For example, in case 2, to see if v interacts with γ, one tests the hypothesis that γ is the same in all groups. If this is rejected, there is an indication of interaction. The largest modification index (Sörbom, 1989) for γ indicates for which group γ is most likely to be different from the others.

If v is a continuous observed variable we assume that v may have a linear effect on α (case 1) or γ (case 2). In case 1 we substitute $a + bv$ for α in Equation 5 to obtain

$$\eta = a + bv + \gamma\xi + \zeta. \tag{8}$$

This is a linear structural equation with one observed and two latent variables. This can be estimated in a straightforward way. In case 2, we substitute $c + dv$ for γ in Equation 5 to obtain

$$\eta = \alpha + c\xi + dv\xi + \zeta. \tag{9}$$

This is a nonlinear structural equation in ξ and v.

If v is a continuous latent variable, we change notation slightly and write ξ_1 for ξ, ξ_2 for v, and ξ_3 for $v\xi$. Equations 8 and 9 then become

$$\eta = a + \gamma\xi_1 + b\xi_2 + \zeta. \tag{10}$$

$$\eta = \alpha + c\xi_1 + d\xi_3 + \zeta. \tag{11}$$

Equations 5–11 are all special cases of the more general form of equation:

$$\eta = \alpha + \gamma_1\xi_1 + \gamma_2\xi_2 + \gamma_3\xi_1\xi_2 + \zeta. \tag{12}$$

This equation was first presented by Kenny and Judd (1984). They had this equation in different notation and with $\eta = y$ and $\alpha = 0$. Jöreskog and Yang (1996) and Yang Jonsson (1997) studied this equation with $\eta = y$. Equation 12 is also used in various chapters of this volume.

Equation 12 is often referred to as an interaction model. This terminology can be explained as follows. Equation 12 can be written as

$$\eta = (\alpha + \gamma_2\xi_2) + (\gamma_1 + \gamma_3\xi_2)\xi_1 + \zeta, \tag{13}$$

or as

$$\eta = (\alpha + \gamma_1\xi_1) + (\gamma_2 + \gamma_3\xi_1)\xi_2 + \zeta, \tag{14}$$

Equation 13 can be interpreted as the regression of η on ξ_1 for given ξ_2 and shows that, if γ_3 is positive, the regression coefficient (slope) increases linearly with ξ_2. So this regression coefficient is larger than γ_1 for those above the mean of ξ_2 and smaller than γ_1 for those below the mean of ξ_2. Similarly, Equation 14 can be interpreted as the regression of η on ξ_2 for given ξ_1 and shows that, if γ_3 is positive, the regression coefficient (slope) increases linearly with ξ_1. So this regression coefficient is larger than γ_2 for those above the mean of ξ_1 and smaller than γ_2 for those below the mean of ξ_1.

ESTIMATION

To be more specific in the following discussion, consider a model where there are three indicators—y_1, y_2, and y_3—of η and two indicators—x_1 and x_2—of ξ_1 and two indicators—x_3 and x_4—of ξ_2. This is similar to the Kenny–Judd model, but they had only one indicator: $y = \eta$. It is exactly the same model used by Yang Jonsson (chap. 2, this volume). The measurement relations in Equations 6 and 7 then take the form:

$$\begin{pmatrix} y_1 \\ y_2 \\ y_3 \end{pmatrix} = \begin{pmatrix} \tau_1^{(y)} \\ \tau_2^{(y)} \\ \tau_3^{(y)} \end{pmatrix} + \begin{pmatrix} 1 \\ \lambda_2^{(y)} \\ \lambda_3^{(y)} \end{pmatrix} \eta + \begin{pmatrix} \varepsilon_1 \\ \varepsilon_2 \\ \varepsilon_3 \end{pmatrix}, \tag{15}$$

$$\begin{pmatrix} x_1 \\ x_2 \\ x_3 \\ x_4 \end{pmatrix} = \begin{pmatrix} \tau_1 \\ \tau_2 \\ \tau_3 \\ \tau_4 \end{pmatrix} + \begin{pmatrix} 1 & 0 & 0 \\ \lambda_2 & 0 & 0 \\ 0 & 1 & 0 \\ 0 & \lambda_4 & 0 \end{pmatrix} \begin{pmatrix} \xi_1 \\ \xi_2 \end{pmatrix} + \begin{pmatrix} \delta_1 \\ \delta_2 \\ \delta_3 \\ \delta_4 \end{pmatrix}. \tag{16}$$

As pointed out by Yang Jonsson (this volume), the parameter α in Equation 12 and the three τs in Equation 15 are not identified. One can add a constant to the three τs and subtract the same constant from α without affecting the mean vector of **y**. Hence, we may set $\alpha = 0$ in Equation 12 and estimate the τs in Equation 15. The estimates of the τs *will not* be equal to the means of the y-variables.

Various procedures have been proposed and discussed in this volume and in earlier literature to estimate the model represented by Equations 12, 6, and 7, or as in the model example 12, 15, and 16. It is desirable to have a procedure that not only provides point estimates of the parameters but also correct standard errors of the parameter estimates. In particular, it is important to estimate γ_3 with a correct standard error so that it can be decided whether γ_3 is zero or not.

The procedures differ with respect to whether the latent and error variables are assumed to be normally distributed, whether observed product variables are used, and between one-step or two-step methods. A further difference between the different approaches is whether they require nonlinear constraints to be imposed.

Product Variables

Jöreskog and Yang (1996) showed that the model cannot be estimated using available SEM programs without an observed product variable as an indicator of the latent product variable $\xi_1\xi_2$. They showed that one product variable is sufficient to identify all the parameters of the model. To estimate the model, one has a choice of using one or more product variables. Kenny and Judd (1984) suggested the use of four product variables. "This seems like a 'natural' choice since there are four 'natural' indicators of $\xi_1\xi_2$, namely x_1x_3, x_1x_4, x_2x_3, x_2x_4" (Yang Jonsson, chap. 2, this volume, p. 20). But because only one product variable is necessary to identify the model, one can choose 1, 2, 3, or 4 product variables.

Is it better to use four product variables or only one? This is discussed by Yang Jonsson (chap. 2, this volume). Yang Jonsson (1997) investigated the differences between a model with one product variable and one with four product variables in a simulation study. The result indicated that bias and mean squared error are smaller for the model with four product variables than for the model with one product variable, but standard errors are more severely underestimated when four product variables are used.

No Product Variables

Two procedures have been proposed that do not require any product indicators. Both procedures assume normality of the latent and error variables.

Contrary to the statement made by Jöreskog and Yang (1996) that the model cannot be estimated using available SEM programs without an observed product variable as an indicator of the latent product variable $\xi_1\xi_2$, Laplante et al. (chap. 9, this volume) claim that the model can be estimated by a simple procedure if τ_y and τ_x in Equations 6 and 7 are zero. Instead, they assume that the means of the latent variables ξ_1 and ξ_2 are free parameters to be estimated. However, this is a very strong assumption that is unlikely to hold in any empirical data. This implies that the mean vector of $\mathbf{x}$ is a linear function of the means of ξ_1 and ξ_2 and the mean vector of $\mathbf{y}$ is linear in the mean of η. In our example, let μ_1, μ_2, μ_3, and μ_4 be the means of x_1, x_2, x_3, and x_4, respectively. Then it follows that $\mu_2/\mu_1 = \lambda_2$ and $\mu_4/\mu_3 = \lambda_4$, where λ_2 and λ_4 are parameters identified by the covariance structure. Obviously, these equations are very unlikely to hold unless x_1 and x_2 are strictly parallel measures, in which case $\lambda_2 = 1$. The same holds for x_3 and x_4 with $\lambda_4 = 1$.

The second approach, which does not make use of any observed product variable, is that of Klein, Moosbrugger, Schermelle-Engel, and Frank (1997) and Schermelle-Engel et al. (chap. 10, this volume). They consider the case of a single indicator $y = \eta$ and suggest using the maximum likelihood method based on the correct joint distribution of y and $\mathbf{x}$. As pointed out by Jöreskog and Yang (1996), if γ_3 is non-zero, this distribution is not multivariate normal even if that of the x-variables is. Assuming normality of ξ_1 and ξ_2 and the error variables in ε and δ, Schermelle-Engel et al. derive this joint distribution of the observables as a continuous mixture of multivariate normal distributions with a standardized normal mixing variable. They claim that this can be approximated by numerical integration using a 16-point Hermite quadrature formula. They have written a special computer program that can be used to estimate the model and demonstrate by simulation that this method works well and gives better standard error than the maximum likelihood method based on normality of the observed variables. Although this procedure presumably is capable of giving consistent and efficient estimates (under the normality assumptions made) with correct asymptotic standard errors, it does not appear to be of practical use for general-purpose nonlinear structural equation modeling.

Two-Step Procedures

Several two-step procedures have been proposed. These do not necessarily assume normality and they are general and flexible enough so that they can be applied to different models and not just to models of the Kenny–Judd type. They are also relatively easy to apply in practice.

Bollen (1995) and Bollen and Paxton (chap. 7, this volume) describe a two-stage least squares (2SLS) procedure using instrumental variables, and Li and Harmer (this volume) present a substantive example of this procedure. In our example, 2SLS involves estimating the relationship

$$y_1 = \alpha + \gamma_1 x_1 + \gamma_2 x_2 + \gamma_3 x_1 x_2 + u, \tag{17}$$

where u is correlated with the variables on the right side in Equation 17 so that ordinary least squares cannot be used to estimate this equation. However, as shown by Bollen and Paxton (chap. 7, this volume), x_3, x_4, and $x_3 x_4$ can be used as instrumental variables. If the model holds, these instrumental variables are uncorrelated with u. The method depends critically on the possibility of finding at least as many instrumental variables as the number of γs on the right side of Equation 17. Bollen and Paxton (chap. 7, this volume) give several examples to show how this can be done. Although the method can be used to estimate the γs in Equation 17 (which are the same as in Equation 12), with their standard errors, it does not estimate all the parameters of the model and therefore it does not give a measure of overall fit of the model. Furthermore, the method does not make use of all the data. Only one of the y-indicators need to be used. However, any of the y-indicators can be used and the estimates of the γs and their standard errors will not be the same for each y-indicator. In the worst case, this can lead to conflicting results. Furthermore, Equation 17 involves only the indicators that are used as reference variables—those that are used to scale the latent variables ξ_1 and ξ_2. However, the choice of reference variables is often rather arbitrary. Do the estimates of the γs and their standard errors depend on which indicators are chosen as reference variables? As the method does not make full use of the data, it would be of interest to compare the standard errors of γs with those of the full information methods proposed by Jöreskog and Yang (1996) and studied by Yang Jonsson (1997).

Ping (1995, 1996a, 1996b) and Ping (chap. 4, this volume) describes another two-step procedure. In this approach, the first step estimates the loadings and error variances of the indicators of the linear latent variables in the measurement model. Then nonlinear indicators of the latent product variable are created as products of the indicators of linear latent variables, and their loadings and error variances are algebraically calculated using the first-step measurement-model estimates. Finally, the nonlinear relationship is estimated within a full structural equation model where the parameters of the measurement model are held fixed at the values calculated in the first step.

This approach also has a number of problems. The observed variables must be normally distributed and errors must be uncorrelated with each other. Although presumably, this approach can yield consistent parameter estimates under the assumptions made, it does not provide correct standard errors of the estimated parameters and correct goodness-of-fit statistics to test the fit of the whole model. The reason for this is the two-step procedure.

To avoid the problem associated with the use of four or more product variables, Ping (1995) suggested using products of single indicators defined as averages over the indicators of each latent variable. This reduces the number of product variables effectively and simplifies the algebra somewhat. In our example, one would form $x_1^* = (x_1 + x_2)/2$ and $x_2^* = (x_3 + x_4)/2$ and then use the single product variable $x_1^* x_2^*$ as a single indicator of $\xi_1\xi_2$. This may be better than using the single indicator x_1x_2, but if the indicators are congeneric (which is the normal case) rather than tau-equivalent, it may produce biased estimates.

Another two-step approach along the same line is proposed by Yang Jonsson (chap. 2, this volume). In the first step, this approach estimates the measurement model for η and ξ assuming these to be freely correlated and individual factor scores for these latent variables. This also provides a test of the measurement model. In the second step, the structural equation model is estimated as if the latent variables were directly observed. This is a simple and straightforward approach that is worth considering. Yang Jonsson (chap. 2, this volume) provides an example. Further studies are needed to compare it with other approaches.

Full-Information Methods

A full-information method is one that fits the moment matrix implied by the model to the corresponding sample moment matrix by minimizing a fit function with respect to all parameters simultaneously. Such methods were proposed by Jöreskog and Yang (1996) and investigated by Yang Jonsson (1997).

In principle, such methods should give the best parameter estimates and standard errors, but the study of Yang Jonsson (1997) indicates that there are several serious problems that make these methods not so useful in practice. These problems have to do with the use of one or more product variables and the associated constraints that these product variables imply. A more important reason for their poor performance in simulation studies, however, is the necessity to use the inverse of an asymptotic covariance matrix in the fit function for WLS or ADF (see Jöreskog & Yang, 1996). This asymptotic covariance matrix can only be estimated accurately from a very large sample; if based on a not-so-large sample, it may be worse using it than not using it. For linear models, the asymptotic covariance matrix is difficult to estimate accurately because it involves fourth-order moments that require large samples to estimate accurately. When product variables are used, this problem becomes even more serious because some of the elements of this matrix are then eighth-order moments, which are even harder to estimate.

A simplified full-information approach follows. Do not assume normality and use only one product variable. This reduces the number of constraints to a minimum because only the constraints that are implied by the use of the product variable are needed. Fit the model by maximum likelihood (ML) method using the augmented moment matrix (see Jöreskog & Yang, 1996). The ML method does not require an asymptotic covariance matrix. The asymptotic covariance matrix is needed, however, to obtain correct standard errors and chi-squares. The point is that it is not used in the iterations (and therefore does not affect the parameter estimates) and it does not have to be inverted. The standard errors and chi-squares produced this way are standard errors and chi-squares for ML estimates corrected for non-normality, sometimes called robust standard errors and chi-squares. (For further information on the implementation of these in LISREL, see the website document www.ssicentral.com/lisrel.newkarl.htm or the topic *Robust Standard Errors and Chi-squares* in the LISREL help file and references given therein.) Jöreskog and Yang (1997) computed these robust standard errors for a model very similar to our model example and found that the standard errors of ML estimates computed under normality grossly underestimate the true standard errors, but the robust standard errors correct most of this bias.

The reason why the augmented moment matrix has to be used rather than the covariance matrix is that the parameters cannot be divided into two mutually exclusive sets of parameters—location and scale—and because of non-normality, the sample mean vector is not uncorrelated with the sample covariance matrix. So the augmented moment matrix (denoted **A**) and the corresponding estimate of the asymptotic covariance matrix of **A** (denoted $\mathbf{W}_a$) must be used in order to obtain correct asymptotic standard errors (see Jöreskog & Yang, 1996; Yang Jonsson, 1997, chap. 2, this volume). The matrix $\mathbf{W}_a$ is singular, but this does not matter because it will not be inverted. The details of this approach are outlined in the Appendix.

CONCLUSION

Consider a typical social-science empirical researcher who has a more or less clear or strong theory suggesting an interaction effect and has collected some data to support this theory. The sample may consist of a few hundred cases. Which of the available procedures should this researcher use?

If the interaction variable is observed and can be used to form some "natural" groups, the multigroup approach is the simplest and most straightforward. If the interaction variable is latent, the factor scores approach or the TSLS approach are probably the most reasonable to use. The use of these will be considerably simplified in the future. In LISREL 9 the factor scores will be obtained along with the estimation of the model and Bollen's (1995) 2SLS estimator can be obtained with one single command in PRELIS 3.

A full-information approach should probably be used only if one has a very large sample and one is capable of understanding how to specify the nonlinear constraints implied by the model.

ACKNOWLEDGMENT

The research reported in this chapter has been supported by the Swedish Council for Research in the Humanities and Social Sciences (HSFR) under the program *Multivariate Statistical Analysis*.

REFERENCES

Agresti, A. (1990). *Categorical data analysis*. New York: Wiley.

Bollen, K. A. (1995). Structural equation models that are nonlinear in latent variables: A least squares estimator. In P. M. Marsden (Ed.), *Sociological methodology 1995*. Cambridge, MA: Blackwell.

Edwards, D. (1995). *Introduction to graphical modeling*. New York: Springer Verlag.
Jöreskog, K. G., & Yang, F. (1996). Nonlinear structural equation models: The Kenny-Judd model with interaction effects. In G. A. Marcoulides & R. E. Schumacker (Eds.), *Advanced structural equation modeling: Issues and techniques* (pp. 57–88). Mahwah, NJ: Lawrence Erlbaum Associates.
Jöreskog, K. G., & Yang, F. (1997). Estimation of interaction models using the augmented moment matrix: Comparison of asymptotic standard errors. In W. Bandilla & F. Faulbaum (Eds.), *SoftStat 97: Advances in statistical software* (pp. 467–478). Stuttgart: Lucius & Lucius.
Kenny, D. A., & Judd, C. M. (1984). Estimating the nonlinear and interactive effects of latent variables. *Psychological Bulletin, 96*, 201–210.
Klein, A., Moosbrugger, H., Schermelle-Engel, K., & Frank, D. (1997). A new approach to the estimation of latent interaction effects in structural equation models. In W. Bandilla & F. Faulbaum (Eds.), *SoftStat 97: Advances in statistical software* (pp. 479–486). Stuttgart: Lucius & Lucius.
McCullagh, P., & Nelder, J. (1989). *Generalized linear models*. London: Chapman & Hall.
Ping, R. A. (1995). A parsimonious estimating technique for interaction and quadratic latent variables. *Journal of Marketing Research, 32*, August, 336–347.
Ping, R. A. (1996a). Latent variable interaction and quadratic effect estimation: A two-step technique using structural equation analysis. *Psychological Bulletin, 119*, 166–175.
Ping, R. A. (1996b). Latent variable regression: A technique for estimating interaction and quadratic coefficients. *Multivariate Behavioral Research, 31*, 95–120.
Sörbom, D. (1989). Model modification. *Psychometrika, 54*, 371–384.
Yang Jonsson, F. (1997). *Non-linear structural equation models: Simulation studies of the Kenny-Judd model*. Unpublished doctoral dissertation, Uppsala University, Sweden.

APPENDIX

For our example model, the LISREL implementation is very similar to that described in the section "LISREL Implementation for WLSA" in chapter 2. It differs from this as follows:

- The last row of Λ_y and Θ_ε are omitted.
- The elements ψ_{42}, ψ_{43}, and ψ_{44} are set free.

Only four nonlinear constraints are needed, namely those defining $\lambda_{85}^{(y)}$, $\theta_{84}^{(\varepsilon)}$, $\theta_{86}^{(\varepsilon)}$, and $\theta_{88}^{(\varepsilon)}$.

Assuming the data on y_1–y_3 and x_1–x_4 are in the file IMDATA.RAW, the following PRELIS input computes the augmented moment matrix **A** and the corresponding asymptotic covariance matrix $\mathbf{W}_a$. These are saved in files `IMDATA.AM`, and `IMDATA.ACA`, respectively.

```
Computing AM, and ACA Matrices
DA NI=7
LA
Y1 Y2 Y3 X1 X2 X3 X4
RA=IMDATA.RAW
CO ALL
```

```
NE X1X2 = X1*X2
OU MA=AM CM=IMDATA.AM AC=IMDATA.ACA
```

To fit and test the model, the following `LISREL` input can be used. The key element here is the `ME=ML` on the last line. Without this, LISREL will use the WLS (ADF) method.

```
Fitting Model by ML using Augmented Moment Matrix
Robust Estimates of Standard Errors and Chi-squares
DA NI=9 NO=number of cases
LA
Y1 Y2 Y3 X1 X2 X3 X4 CONST
CM=IMDATA.AM
AC=IMDATA.ACA
MO NY=8 NE=5 NX=1 BE=FU GA=FI TE=SY PS=SY,FI FI
FR LY(2,1) LY(3,1) LY(5,2) LY(7,3) BE(1,2) BE(1,3) BE(1,4)
FR LY(1,5) LY(2,5) LY(3,5) LY(4,5) LY(5,5) LY(6,5) LY(7,5)
FR PS(1,1) PS(2,2) PS(3,2) PS(3,3) PS(4,2) PS(4,3) PS(4,4)
VA 1 LY(1,1) LY(4,2) LY(6,3) LY(8,4) GA(5)
EQ LY(8,2) LY(6,5)
EQ LY(8,3) LY(4,5)
CO LY(8,5)=LY(4,5)*LY(6,5)
EQ GA(4) PS(3,2)
CO TE(8,4)=LY(6,5)*TE(4,4)
CO TE(8,6)=LY(4,5)*TE(6,6)
CO TE(8,8)=LY(4,5)**2*TE(6,6)+LY(6,5)**2*TE(4,4)+PS(2,2)*TE(6,6)+ C
PS(3,3)*TE(4,4)+TE(4,4)*TE(6,6)
OU ME=ML AD=OFF
```

Author Index

V

W

Y

Z

Subject Index

G

I

K

L

M

N

T

U

W

About the Authors

Kenneth A. Bollen is the Zachary Smith Professor of Sociology at the University of North Carolina at Chapel Hill and is a Fellow of the Carolina Population Center at UNC. Structural equation modeling (SEM) is his primary area of statistical research. He is applying SEM in the analysis of access and quality of family planning facilities in developing countries. In addition, he continues to do research in the comparative study of liberal democracies. He is the author of *Structural Equations With Latent Variables* (Wiley) and co-editor (with J. S. Long) of *Testing Structural Equation Models* (Sage). Recent papers have been published in *American Sociological Review*, *Sociological Methodology*, *Psychometrika*, and *Demography*.
(Email: bollen@gibbs.oit.unc.edu)

Louis-Georges Cournoyer received his Ph.D. in Psychology from Université de Montréal in 1995. He then became Postdoctoral Fellow in Psychiatry at McGill University and in Criminology at Université de Montréal. His current research focuses on the therapeutic processes and outcomes of drug abuse treatments.

Darin J. Erikson is a project coordinator at the Women's Health Sciences Division of the National Center for PTSD. His primary interest in structural equation modeling is its application to the analysis of longitudinal data.

Peter Harmer is an Associate Professor in the Department of Exercise Science-Sports Medicine at Willamette University in Salem, Oregon. He received his Ph.D. in Human Movement Studies from the University of Oregon and has published work in a variety of disciplines, including athletic medicine, exercise physiology, measurement, sport psychology, and philosophy.

Fan Yang Jonsson received her Ph.D in Statistics at Uppsala University in 1997. She is currently a Research Associate in the Department of Statistics at Uppsala University, working with Karl Jöreskog. Her research interests are in the areas of measurement, factor analysis, and structural equation modeling, especially nonlinear structural equation modeling.
(Email: fan.yang@statistiks.uu.se)

Karl G. Jöreskog is Professor of Multivariate Statistical Analysis at Uppsala University, Sweden. His main interests are in the theory and applications of structural equation models and other types of multivariate analysis, particularly their applications in the social and behavioral sciences. He is coauthor of *LISREL 7—A Guide to the Program and Applications*, published by SPSS in 1989, and *LISREL 8—Structural Equation Modeling With the SIMPLIS Command Language*, published by SSI in 1993. He also coauthored with Fan Yang Jonsson, chapter 3: "Nonlinear Structural Equation Models: The Kenny-Judd Model With Interaction Effects," In G. A. Marcoulides and R. E. Schumacker (Eds.), *Advanced Structural Equation Modeling: Issues and Techniques,* published by Lawrence Erlbaum Associates, Publishers Inc., Mahwah, NJ, in 1996.
(Email: karl.joreskog@statistiks.uu.se)

Andreas Klein received his Diploma in Mathematics at Goethe-University, Frankfurt am Main, Germany, where he is a research associate at the Department of Psychological Methodology. His current research interests include structural equation modeling, research methods, and multivariate statistics. At present, he is finishing his doctoral thesis in Mathematical Psychology, which focuses on the development of estimation methods for nonlinear structural equation models. He has presented his results at several international and national conferences, and is a member of the German Psychological Association.
(Email: A.Klein@psych.uni-frankfurt.de)

Benoît LaPlante is Associate Professor at INRS-Culture et Société, a research center that is part of Université de Québec. He received his Ph.D. in Sociology from Université de Montréal. His current research deals with artistic occupations and labor in the cultural sector as well as quantitative methodology issues, mainly structural equation modeling and event history analysis.
(Email: benoit.laplante@ehess.fr)

Fuzhong Li received his Ph.D. from Oregon State University and is currently affiliated with the Oregon Research Institute. His research interests include

exercise health and motivation and statistical methodology, especially structural equation modeling.
(Email: fuzhongL@ori.org)

Edward E. Rigdon is an Associate Professor of Marketing at Georgia State University in Atlanta. Dr. Rigdon teaches a doctoral seminar in structural equation modeling (SEM), serves as methodological specialist on dissertation committees, and consults with commercial researchers on SEM problems. Dr. Rigdon's SEM research has been published in the *Journal of Marketing Research, Multivariate Behavioral Research, Structural Equation Modeling*, and in two books. Dr. Rigdon serves on the editorial board of *Structural Equation Modeling*. He is a cofounder of SEMNET, an Internet-based discussion list devoted to SEM, and he maintains a World Wide Web site devoted to SEM.
(Email: mkteer@langate.gsu.edu)

George A. Marcoulides is Professor of Statistics at California State University at Fullerton and Adjunct Professor at the University of California at Irvine. He is the recipient of the 1991 UCEA William J. Davis Memorial Award for outstanding scholarship. He is currently the Editor of the Quantitative Methodology Book Series, Editor of the *Structural Equation Modeling* journal, and on the editorial board of several other measurement and statistics journals. His research interests include generalizability theory and structural equation modeling.
(Email: gmarcoulides@fullerton.edu)

Dr. Helfried Moosbrugger is Professor at Goethe-University, Frankfurt am Main, Germany, and head of the Department of Psychological Methodology, where he teaches psychological statistics, research methods, test theory, and psychological assessment. He received his Ph.D. from the University of Innsbruck, Austria, and has published about 100 articles, books and chapters, and has presented at numerous conferences in Europe and the United States. At present, his research interests are moderated structural equation modeling and computer-based psychological assessment. Dr. Moosbrugger is vice-president of the Methodology Section of the German Psychological Association, serves on the editorial board of Methods of Psychological Research-Online, and is a member of the International Society for the Study of Individual Differences.

Michael C. Neale is Associate Professor of Psychiatry and Psychology at Virginia Commonwealth University. He has over 125 published articles, books, and chapters, and is the author of Mx, a package for structural equation and other mathematical modeling that combines a matrix algebra interpreter with optimization software. His work concentrates on developing

statistical methods to attack problems in psychiatric genetics, such as comorbidity, age at onset, heterogeneity, and the linkage analysis of complex phenotypes. These methods are applied by his colleagues at VCU and elsewhere to a variety of behavioral and physical traits.
(Email: neale@ruby.vcu.edu)

Pamela Paxton is a doctoral candidate in sociology at the University of North Carolina at Chapel Hill, with research interests in quantitative methods, political sociology, and stratification. Her dissertation empirically investigates the relationship between civil society and democracy. Some of her recent publications have appeared in *Social Science Research* and *American Sociological Review*.

Robert A. Ping, Jr., an Associate Professor at Wright State University in Dayton, Ohio, received his Ph.D. in Marketing from the University of Cincinnati. With publications and acceptances in the *Psychological Bulletin*, the *Journal of Marketing Research*, *Multivariate Behavioral Research*, the *Journal of Retailing*, and elsewhere, his research concerns long-term economic/social exchange relationships and nonlinear estimation in latent variables. A Fellow in The Academy of Marketing Science, Professor Ping is an ad hoc reviewer for several journals, including the *Journal of Marketing*, the *Journal of Personality and Social Psychology*, *Multivariate Behavioral Research*, the *Journal of Retailing*, and the *Journal of the Academy of Marketing Science*.
(Email: rping@desire.wright.edu)

Stéphane Sabourin is Professor of Psychology at Université Laval in Québec City. His research and clinical practice deal with couple distress.

Karin Schermelleh-Engel received her Ph.D. from Goethe-University, Frankfurt am Main, Germany, where she is a lecturer and research associate in the Department of Psychological Methodology. She has published several articles and book chapters on methodology as well as on chronic pain and pain assessment, and has developed the Pain Regulation Questionnaire. Her current research interests include nonlinear structural equation modeling, research methods, and psychological assessment. She has presented her results at several international and national conferences, and is a member of the German Psychological Association.
(Email: schermelleh-Engel@psych.uni-frankfurt.de)

Randall E. Schumacker is Professor of Educational Research at the University of North Texas and Professor of Medical Education at the University of North Texas Health Science Center. He was a 1996 recepient of an outstanding scholar award at UNT, has taught SEM workshops around the country, and teaches a doctoral seminar in structural equation modeling at

UNT. Dr. Schumacker was the founder and past editor of *Structural Equation Modeling: A Multidisciplinary Journal* (1994–1998), coeditor of *Advanced Structural Equation Modeling: Issues and Techniques* (1996) with Dr. George A. Marcoulides, and coauthor of *A Beginner's Guide to Structural Equation Modeling* (1996) with Dr. Richard G. Lomax. He currently serves on the editoral board of several journals and pursues his research interests in structural equation modeling, Rasch measurement, and medical education.
(Email: rschumacker@unt.edu)

Richard L. Tate is an Associate Professor in the Department of Educational Research, College of Education, at Florida State University. He specializes in applied statistics and measurement with current research interests in item response theory and structural equation modeling.
(email: rtate@garnet.acns.fsu.edu)

Phillip K. Wood received his Ph.D. from the University of Minnesota in 1985. He is currently an Associate Professor at the University of Missouri–Columbia. His research interests include structural equation modeling, particularly as it is related to the assessment of differential change and growth. Substantive research interests include the assessment of adult cognitive development, college outcomes, and graduate training in statistics.
(Email: wood@psysparc.psyc.missouri.edu)

Werner Wothke is President of SmallWaters Corporation. He received his Ph.D. in methodology of behavioral sciences from the University of Chicago in 1984. He has served as consultant to statistical reporting and software publishing companies and has spoken at universities and professional societies. In 1993, Dr. Wothke cofounded SmallWaters Corporation as a venue for publishing and promoting innovative statistical software. His research interests are in multivariate model building and statistical computing.
(Email: wewo@smallwaters.com)

John Wright is Professor of Psychology at Université de Montréal. He obtained a Ph.D. in Clinical Psychology from the University of Wisconsin, Madison. His research and clinical practice deal with couple distress and child sexual abuse.